Maria Virvou and Lakhmi C. Jain (Eds.)

Intelligent Interactive Systems in Knowledge-Based Environments

Studies in Computational Intelligence, Volume 104

Editor-in-chief
Prof. Janusz Kacprzyk
Systems Research Institute
Polish Academy of Sciences
ul. Newelska 6
01-447 Warsaw
Poland
E-mail: kacprzyk@ibspan.waw.pl

Further volumes of this series can be found on our homepage: springer.com

Vol. 80. Joachim Diederich
Rule Extraction from Support Vector Machines, 2008
ISBN 978-3-540-75389-6

Vol. 81. K. Sridharan
Robotic Exploration and Landmark Determination, 2008
ISBN 978-3-540-75393-3

Vol. 82. Ajith Abraham, Crina Grosan and Witold Pedrycz (Eds.)
Engineering Evolutionary Intelligent Systems, 2008
ISBN 978-3-540-75395-7

Vol. 83. Bhanu Prasad and S.R.M. Prasanna (Eds.)
Speech, Audio, Image and Biomedical Signal Processing using Neural Networks, 2008
ISBN 978-3-540-75397-1

Vol. 84. Marek R. Ogiela and Ryszard Tadeusiewicz
Modern Computational Intelligence Methods for the Interpretation of Medical Images, 2008
ISBN 978-3-540-75399-5

Vol. 85. Arpad Kelemen, Ajith Abraham and Yulan Liang (Eds.)
Computational Intelligence in Medical Informatics, 2008
ISBN 978-3-540-75766-5

Vol. 86. Zbigniew Les and Mogdalena Les
Shape Understanding Systems, 2008
ISBN 978-3-540-75768-9

Vol. 87. Yuri Avramenko and Andrzej Kraslawski
Case Based Design, 2008
ISBN 978-3-540-75705-4

Vol. 88. Tina Yu, David Davis, Cem Baydar and Rajkumar Roy (Eds.)
Evolutionary Computation in Practice, 2008
ISBN 978-3-540-75770-2

Vol. 89. Ito Takayuki, Hattori Hiromitsu, Zhang Minjie and Matsuo Tokuro (Eds.)
Rational, Robust, Secure, 2008
ISBN 978-3-540-76281-2

Vol. 90. Simone Marinai and Hiromichi Fujisawa (Eds.)
Machine Learning in Document Analysis and Recognition, 2008
ISBN 978-3-540-76279-9

Vol. 91. Horst Bunke, Kandel Abraham and Last Mark (Eds.)
Applied Pattern Recognition, 2008
ISBN 978-3-540-76830-2

Vol. 92. Ang Yang, Yin Shan and Lam Thu Bui (Eds.)
Success in Evolutionary Computation, 2008
ISBN 978-3-540-76285-0

Vol. 93. Manolis Wallace, Marios Angelides and Phivos Mylonas (Eds.)
Advances in Semantic Media Adaptation and Personalization, 2008
ISBN 978-3-540-76359-8

Vol. 94. Arpad Kelemen, Ajith Abraham and Yuehui Chen (Eds.)
Computational Intelligence in Bioinformatics, 2008
ISBN 978-3-540-76802-9

Vol. 95. Radu Dogaru
Systematic Design for Emergence in Cellular Nonlinear Networks, 2008
ISBN 978-3-540-76800-5

Vol. 96. Aboul-Ella Hassanien, Ajith Abraham and Janusz Kacprzyk (Eds.)
Computational Intelligence in Multimedia Processing: Recent Advances, 2008
ISBN 978-3-540-76826-5

Vol. 97. Gloria Phillips-Wren, Nikhil Ichalkaranje and Lakhmi C. Jain (Eds.)
Intelligent Decision Making: An AI-Based Approach, 2008
ISBN 978-3-540-76829-9

Vol. 98. Ashish Ghosh, Satchidananda Dehuri and Susmita Ghosh (Eds.)
Multi-Objective Evolutionary Algorithms for Knowledge Discovery from Databases, 2008
ISBN 978-3-540-77466-2

Vol. 99. George Meghabghab and Abraham Kandel
Search Engines, Link Analysis, and User's Web Behavior, 2008
ISBN 978-3-540-77468-6

Vol. 100. Anthony Brabazon and Michael O'Neill (Eds.)
Natural Computing in Computational Finance, 2008
ISBN 978-3-540-77476-1

Vol. 101. Michael Granitzer, Mathias Lux and Marc Spaniol (Eds.)
Multimedia Semantics - The Role of Metadata, 2008
ISBN 978-3-540-77472-3

Vol. 102. Carlos Cotta, Simeon Reich, Robert Schaefer and Antoni Ligeza (Eds.)
Knowledge-Driven Computing, 2008
ISBN 978-3-540-77474-7

Vol. 103. Devendra K. Chaturvedi
Soft Computing Techniques and its Applications in Electrical Engineering, 2008
ISBN 978-3-540-77480-8

Vol. 104. Maria Virvou and Lakhmi C. Jain (Eds.)
Intelligent Interactive Systems in Knowledge-Based Environments, 2008
ISBN 978-3-540-77470-9

Maria Virvou
Lakhmi C. Jain (Eds.)

Intelligent Interactive Systems in Knowledge-Based Environments

With 67 Figures and 40 Tables

Prof. Maria Virvou
Department of Informatics
University of Piraeus
Karaoli & Dimitriou Street 80
185 34 Piraeus
Greece
mvirvou@unipi.gr

Prof. Lakhmi C. Jain
School of Electrical & Information Engineering
University of South Australia
KES Centre
Mawson Lakes Campus
Adelaide SA 5095
Australia
lakhmi.jain@unisa.edu.au

ISBN 978-3-540-77470-9 e-ISBN 978-3-540-77471-6

Studies in Computational Intelligence ISSN 1860-949X

Library of Congress Control Number: 2008920948

Cover design: Deblik, Berlin, Germany

Printed on acid-free paper

9 8 7 6 5 4 3 2 1

springer.com

Preface

Tremendous advances in intelligent paradigms, such as artificial neural systems, machine learning, evolutionary computing, web-based systems and intelligent agent systems, have immensely contributed in the field of interactive systems. The personalization of intelligent interactive systems has become reality. Researchers are working on fusing adaptivity and learning in interactive systems. Intelligent recommender systems, intelligent tutoring systems, intelligent web-shops, intelligent interactive TV, intelligent mobile services, affective systems and narrative environments constitute typical intelligent interactive systems. However, the list of intelligent interactive systems is not exhausted to the aforementioned applications, as research and developments in the area expand to many more applications.

The main aim of this book is to report a sample of the most recent advances in the field of intelligent interactive systems in knowledge-based environments. This book consists of ten chapters reflecting the theoretical foundations as well as typical applications in the area of intelligent interactive systems.

We wish to express our gratitude to the authors and reviewers for their wonderful contributions. Thanks are due to Springer-Verlag for their editorial support. We would also like to express our sincere thanks to Ms Sridevi Ravi for her wonderful editorial support. We believe that this book would help in creating interest among researchers and practitioners towards realizing human-like machines. This book would prove useful to the researchers, professors, research students and practitioners as it reports novel research work on challenging topics in the area of intelligent interactive systems. Moreover, special emphasis has been put on highlighting issues concerning the development process of such complex interactive systems, thus revisiting the difficult issue of knowledge engineering of such systems. In this way, the book aims at providing the readers with a better understanding of how intelligent interactive systems can be successfully implemented to incorporate recent trends and advances in theory and applications of intelligent systems.

Greece *Maria Virvou*
Australia *Lakhmi C. Jain*

Contents

1

Intelligent Interactive Systems in Knowledge-Based Environments: An Introduction

Maria Virvou[1] and Lakhmi C. Jain[2]

[1] Department of Informatics, University of Piraeus, Piraeus, Greece
[2] School of Electrical & Information Engineering, University of South Australia, Adelaide, Australia

Summary. This chapter presents an introduction to the important topic of intelligent interactive systems in knowledge-based environments. Recent advancements and new trends in computers, such as Web-based technology, mobile software, integrated interactive environments have rendered computers accessible and indispensable to many users. However, at the same time new challenges have been introduced to the already complex task of knowledge-engineering with regard to Intelligent Interactive Software (IIS). IISs have to be personalized to address individual needs of users, they have to ensure interoperability, succeed in achieving reusability, possibly extend to further modalities that include vision and speech, provide emotional intelligence but most of all they have to retain usability and utility to justify for the complexity of their development. Given the fact that there are many underlying theories that can be applied in IIS (cognitive theories, decision making theories, clustering algorithms, neural networks and so on.) the development process is different for each case and in any case difficult. Achieving usability and utility of the end-product should be among the most important goals. Thus the development process has to take into account real-life context such as case studies about users' requirements, experts' analyses, human observers' opinions in several cycles of executable releases of IISs before an IIS may be delivered to its end-users.

1.1 Introduction

At a time when computers are more widespread than ever and the Web becomes indispensable for everyday activities, Intelligent Interactive Systems (IIS) are also needed more than ever. The solution of "one-fits-all" is no longer applicable to wide ranges of users of various backgrounds and needs. Therefore one important goal of many intelligent interactive systems is dynamic personalization and adaptivity to users. Intelligent recommender systems, intelligent tutoring systems, intelligent web-stores, intelligent interactive TV, intelligent mobile services constitute a set of typical examples of intelligent interactive

M. Virvou and L.C. Jain: *Intelligent Interactive Systems in Knowledge-Based Environments: An Introduction*, Studies in Computational Intelligence (SCI) **104**, 1–8 (2008)
www.springerlink.com

systems. However, the list of application domains of intelligent interactive systems is not limited to a few examples.

In the case of the Web, which achieves interoperability, platform independence and easy access of computer facilities to users, user-friendliness and dynamic adaptation to the needs of individual users is still a great challenge. The personalization of services offered by a Web site is an important step in the direction of alleviating information overload and hence creating trustworthy relationships between the Web site and the visitor customer [25].

However, to create trustworthy relationships with the users, there is a compelling need to have evaluated the effects of personalized intelligent interactive systems and have accurately specified the requirements of such systems in the first place. Knowledge engineering techniques include protocol analysis, observations, interviews and introspection, case analysis and questionnaires [3]. However, often the result is a "data dump" that leads to knowledge that is poorly structured, inflexible and incomplete [32]. After an exhaustive review of the literature of Intelligent User Interfaces [7], it is concluded that there is a shortage of guidelines available for the development of such applications.

It becomes apparent that the problem of knowledge engineering of intelligent interactive systems has to be revisited. Presently, this is even more the case when new challenges concerning the design of intelligent interactive systems have to be met.

1.2 Intelligent Interactive Systems

Traditionally, software developers address the problem of the differences in the background of software users by developing customizable software. Customizability is the modifiability of the user interface by the user or the system; however there is a distinction between the user-initiated and system-initiated modification, referring to the former as adaptability and the latter as adaptivity [8]. In any case customizable software always incorporates some kind of user modeling techniques.

As reported in [28], most computer system users have had some experience with user modeling through systems that let users modify system parameters or otherwise customize the system's interaction behavior to their own preferences. However, in the case of adaptability, which refers to the user's ability to adjust the form of input and output, this customization may be very limited. In many cases, the user is only allowed to adjust the position of soft buttons on the screen or redefine command names.

User profiles are also commonly employed to add adaptability to programs that are used by a wide range of individuals, such as word processors, text editors or electronic mail systems. In these cases, the user usually creates or modifies an initialization file, or answers a series of questions to establish initial values for the modifiable system parameters. On subsequent program invocations these parameter settings are retrieved, so the user can customize the system's interactions to his or her own taste [17].

However, adaptability is not flexible enough to provide real help to software users. First, adjustments made by users are rather static, meaning that they do not change once determined. Second and most importantly, in the case of adaptability users have to know how to customize software to their particular needs. As pointed out in [17], it is unreasonable to expect users to build detailed profiles of their domain beliefs because that effort would take much longer than the primary interaction itself. On the other hand, adaptivity provides automatic customization of software to the users' needs based on more sophisticated user modeling techniques. It may also provide intelligent help, which is more flexible than traditional on-line help. Adaptivity is automatic customization of the user interface by the system [4]. Decisions for adaptation can be based on user expertise or observed repetition of certain tasks. A system may be trained to recognize the behavior of an expert or novice user based on individual user models. Then, it may adjust its dialogue control or help system automatically to match the needs of the current user.

1.3 Current Applications of Intelligent Interactive Systems

Applications of intelligent interactive systems span many areas and impose many requirements. In many cases, in intelligent interactive systems, intelligent agents are used, either over the Web or in standalone applications where user interactions are particularly demanding. As pointed out in [2], design of user interaction of web-based agent systems necessitates new approaches in relation to control, task allocation, transparency and user's privacy protection. Web-based technology needs more intelligence for Web-based user interfaces to be more user-friendly. Hypertext systems are quite popular for the design of Web-based user interfaces. Hypertext stores text in a network and connects related sections of text using selectable links. By clicking on a link, the user can go to a related subject instantly. However, a common problem to hypertext systems is navigation; the user can get lost within the hypertext and lose track of where s/he is and where s/he has been [4]. Thus, hypertext constitutes a common area of application of adaptive and intelligent techniques in Web-based technology with the use of user models.

There are many more challenges to be met when intelligent behavior of interactive systems is required in several other topics that have recently attracted researchers' attention. For example, the area of multi-modal user interfaces constitutes a topic that develops quickly but still needs many more research efforts. Among many functionalities, multi-modal interfaces are being used for augmented reality user interfaces (e.g., [14]), or they are used to provide better recognition of human faces (e.g., [29, 38]) and gestures (e.g., [22] or they are used in combinations of more topics of research in intelligent interactive systems such as personalization and mobile devices (e.g., [1]).

Recently, intelligent interactive systems seem to be needed for mobile devices as well. Again, the domain of these applications may vary quite a lot. For example, in [31], the authors point out that the study of intelligent user interfaces and user modeling and adaptation is well suited for augmenting educational visits to museums. In another very different case [19], a distributed architecture based on the usage of intelligent user interfaces and multi-agent systems is proposed to facilitate cooperative internet-based remote interaction with a multi-robot system. Intelligent tutoring systems are also using mobile features, intelligent e-commerce applications are eventually turned to m-commerce since they aim at incorporating mobile features and so on.

At the same time, research on interactive and narrative technologies has grown for the purpose of creating software that is more motivating and engaging for the users than conventional software. Querec and Chevalier [26] note that virtual reality allows users to immerse in a universe where the physical environment and human actors' behavior are simulated.

A very important application area of intelligent interactive systems is the area of education. In this area, appropriate user interfaces are needed to handle the difficulty in communicating knowledge to students through computers. User modeling is frequently used to provide adaptivity and intelligence to such systems, so that instruction may be dynamically adapted to individual students that learn while using a computer. The focus of these systems is often placed in complex topics that vary quite a lot. For example, Kerly et al. in [18] aim at constructing natural language negotiation of open learner models, whereas in the case of EpiList, the focus is on the implicit development of generic cognitive skills [13] or in other cases the emphasis is put on using artificial intelligence for improving teaching [27].

Determining users' feelings while they interact with computers has recently become a very important issue in the area of intelligent interactive systems. In addition to that another very important issue is creating computer responses that are appropriate to users' feelings. There is a lot of ongoing research on these issues and many different approaches are reported in the literature (e.g., [5, 10, 16]).

1.4 Development Process of Intelligent Interactive Systems

Knowledge elicitation and acquisition from domain experts are needed for constructing the framework of the knowledge bases of the intelligent interactive systems. However, it is widely acknowledged that the process of knowledge acquisition is a bottleneck in the development of expert systems [13–16].

The inclusion of end-users and human domain experts in the development process of intelligent interactive systems is increasingly needed [34] to create trustworthy and effective systems. These processes are difficult and time consuming. Thus, it is a drawback of current intelligent systems when they are always built from scratch [21] without using any already constructed components or at least some kind of prior knowledge on the development processes [33]. For example, in the area of Intelligent Tutoring Systems (ITS), which is an area of IIS, Woolf and Cunningham [37] have estimated that an hour of instructional material requires more than 200 h of ITS development time. This shows that enhancing the development processes is an important step forward towards the achievement of effective Intelligent Interactive Systems [23].

Common approaches proposed in the literature for incorporating intelligence in user interfaces include probabilistic reasoning through Bayesian Networks, machine-learning algorithms, neural networks, case-Based Reasoning, cognitive reasoning or decision-making theories and so on. Incorporating such theories into intelligent interactive systems constitutes a very difficult process that requires many experimental studies for various purposes depending on the theory, algorithm or technique used. For example, the success of a neural network-based IIS relies heavily on the experimental data that has been used for training the neural network. On the other hand, in the case of an IIS that has been based on a multi-criteria decision making theory, experimental studies may play an important role on the specification of criteria and their weights [15], which is also a crucial point for the success of the whole IIS.

Moreover, a review [6] on personalization and user adapted interaction points out that there are insufficient empirical evaluations. Du Boulay [10] points out that one quite established criterion of effectiveness of an intelligent system is whether the intelligent version works better than the non intelligent version of the same system. However, to achieve this reasonable comparison of two versions of a system as an evaluation, a knowledge engineer should have designed two versions of the same system in early stages of the development process. Again, the development process of an IIS is a crucial point for its validity and usefulness.

1.5 Chapters Included in This Book

This book includes ten chapters. Chapter one introduces intelligent interactive systems as well as provides a brief introduction to the chapters included in this book.

Chapter two by Alepis, et al. is on requirements analysis and design of an affective bi-modal intelligent tutoring system that recognizes student's emotions based on their words and actions that are identified by the microphone and keyboard. The system uses an innovative approach that combines evidence from the two modes of interaction using a multi-criteria decision making theory.

Chapter three by Bibi and Stamelos is on estimating the development cost for intelligent systems. The machine learning techniques considered by the authors are analogy-based estimation, classification trees, rule induction and Bayesian belief networks. The results of this research are summarized in the end.

Chapter four by Brna is on narrative interactive learning environments. The author has described various notions of narratives for enhancing learning.

Chapter five by Felfernig is on knowledge acquisition for configurable products and services. Two innovative approaches of knowledge acquisition for configurable products and services are presented.

Chapter six by O'Grady et al. is on interaction modalities in mobile contexts. This chapter presents the importance of interaction and its implementation in the context of mobile human-computer interaction.

Chapter seven by Stathopoulou and Tsihrintzis is on automated processing and classification of face images for human-computer interaction applications. The authors have reported a novel system for processing multiple camera images of faces for identifying facial features.

Chapter eight by Hadzilacos, et al. is on modeling distance learning based on assignment and examination data. The authors have used intelligent paradigms such as decision trees and genetic algorithms to analyze the academic performance of students for selected courses.

Chapter nine by Masthoff is on group adaptation and group modeling. The author has presented a system which adapts to a group of users.

The final chapter by Paliouras et al. is on a personalized news aggregator on the Web. The system aggregates news from various electronic news publishers and distributors and organizes them according to pre-defined categories and constructs news via a Web-based interface. The authors have presented the results of a user study.

1.6 Summary

This chapter has presented a brief overview of intelligent interactive systems in knowledge-based environments. A sample of recent research directions in interactive systems undertaken by the researchers is reported in this book.

References

1. Aoidh, E.M. Personalised multimodal interfaces for mobile geographic information systems, *Lecture Notes in Computer Science* (including subseries Lecture Notes in Artificial Intelligence and Lecture Notes in Bioinformatics) 4018 LNCS, pp. 452–456 (2006)
2. Avouris, N.M., Solomos, K.G. User interaction with web-based agents for distance learning, Australian Computer Journal 33(1), 16–29 (2001)
3. Bell, J., Hardiman, R.J. The third role – the naturalistic knowledge engineer. In Diaper, D. (ed.): *Knowledge Elicitation: Principles, Techniques, and Applications*, Chichester, England, Ellis Horwood Ltd., pp. 49–85 (1989)

4. Boose, J.H. A survey of knowledge acquisition techniques and tools. In Buchanan, B.G., Wilkins, D.C. (eds.): *Readings in Knowledge Acquisition and Learning*, San Mateo, CA, Morgan Kaufmann, 29–56 (1993)
5. Catucci, G., Abbattista, F., Gadaleta, R.C., Guaccero, D., Semeraro, G. Empathy: A computational framework for emotion generation, Advances in Soft Computing 2006, 265–277 (2006)
6. Chin, D.N. Empirical evaluation of user models and user-adapted systems, User Modeling and User Adapted Interaction 11(1–2), 181–194 (2001)
7. Delisle, S., Moulin B. User interfaces and help systems: from helplessness to intelligent assistance, Artificial Intelligence Review 18(2), 117–157 (2002)
8. Dix, A., Finlay, J., Abowd, G. Beale, R. *Human Computer Interaction*, Prentice Hall, New York (1993)
9. D'Mello, S., Graesser, A. Affect detection from human-computer dialogue with an intelligent tutoring system, *Lecture Notes in Computer Science* (including subseries Lecture Notes in Artificial Intelligence and Lecture Notes in Bioinformatics) 4133 LNAI, pp. 54–67 (2006)
10. du Boulay, B. Can we learn from ITSs? In G. Gautier, C. Frasson, K. VanLehn (eds.): *ITS 2000, Lecture Notes in Computer Science*, Vol. 1839, pp. 9–17 (2000)
11. Garg-Janardan, C., Salvendy, G. A structured knowledge elicitation methodology for building expert systems. International Journal of Man-Machine Studies 29(4), 377–406 (1988)
12. Glassner, A. Interactive storytelling: people, stories and games. In O. Balet, G. Subsol, P. Torguet (eds.): *International Conference on Virtual Storytelling* (ICVS) 2001, LNCS 2197, pp. 51–60, Springer, Berlin Heidelberg New York (2001)
13. Goh, G.M., Quek, C. EpiList: An intelligent tutoring system shell for implicit development of generic cognitive skills that support bottom-up knowledge construction, IEEE Transactions on Systems, Man, and Cybernetics Part A: Systems and Humans 37(1), 58–71 (2007)
14. Hilliges, O., Sandor, C., Klinker, G. Interactive prototyping for ubiquitous augmented reality user interfaces, International Conference on Intelligent User Interfaces, Proceedings IUI 2006, 285–287 (2006)
15. Kabassi, K., Virvou, M. A knowledge-based software life-cycle framework for the incorporation of multicriteria analysis in intelligent user interfaces, IEEE Transactions on Knowledge and Data Engineering 18(9), 1265–1277 (2006)
16. Kapoor, A., Burleson, W., Picard, R.W. Automatic prediction of frustration, International Journal of Human Computer Studies 65(8), 724–736 (2007)
17. Kass, R. Building a user model implicitly from a cooperative advisory dialog. User Modeling and User-Adapted Interaction 1, 203–258 (1991)
18. Kerly, A., Hall, P., Bull, S. Bringing chatbots into education: towards natural language negotiation of open learner models, Knowledge-Based Systems 20(2), 177–185 (2007)
19. Khamis, A., Abdel-Rahman, A., Kamel, M. A distributed architecture for mobile multirobot remote interaction, Proceedings – DIS 2006: IEEE Workshop on Distributed Intelligent Systems – Collective Intelligence and Its Applications 2006, 61–66 (2006)
20. Kitto, C.M., Boose, J.H. Selecting knowledge acquisition tools and strategies based on application characteristics. International Journal of Man-Machine Studies 31(2), 149–160 (1989)

21. Mizoguchi, R., Bourdeau, J. Using ontological engineering to overcome common AI-ED problems, International Journal AIED, 11(2), 107–121 (2000)
22. Morency, L.-P., Darrell, T. Head gesture recognition in intelligent interfaces: the role of context in improving recognition, International Conference on Intelligent User Interfaces, Proceedings IUI 2006, 32–38 (2006)
23. Moundridou, M., Virvou, M. Analysis and design of a Web-based authoring tool generating intelligent tutoring systems, Computers and Education 40(2), 157–181 (2003)
24. Oard, D.W. The state of the art in text filtering, User Modeling and User-Adapted Interaction 7(3), 141–178 (1997)
25. Pierrakos, D., Paliouras, G., Papatheodorou, C., Spyropoulos D.: Web usage mining as a tool for personalization: a survey, User Modeling and User Adapted Interaction 13, 311–372 (2003)
26. Querrec, R., Chevaillier, P. Virtual storytelling for training: an application to fire fighting in industrial environment. In O. Balet, G. Subsol, P. Torguet (eds.): International Conference on Virtual Storytelling (ICVS) 2001, LNCS 2197, pp. 201–204, Springer, Berlin Heidelberg New York (2001)
27. Reis, M.M., Paladini, E.P., Khator, S., Sommer, W.A. Artificial intelligence approach to support statistical quality control teaching, Computers and Education 47(4), 448–464 (2006)
28. Rich, E. Users as individuals: individualizing user models, International Journal of Man-Machine Studies, 18, 199–214 (1983)
29. Sirlantzis, K., Hoque, S., Fairhurst, M.C. Diversity in multiple classifier ensembles based on binary feature quantisation with application to face recognition, Applied Soft Computing Journal 8(1), 437–445 (2008)
30. Stevens, C. Automating the creation of information filters, Communications of the ACM, 35(12), 48 (1992)
31. Stock, O. Zancanaro, M, Busetta, P., Callaway, C., Krüger, A., Kruppa, M., Kuflik, T., Not, E., Rocchi, C. Adaptive, intelligent presentation of information for the museum visitor in PEACH, User Modelling and User-Adapted Interaction, 17(3), 257–304 (2007)
32. Thurman, D.A., Brann, D.M., Mitchell, C.M. An architecture to support incremental automation of complex systems. In Proceedings of the 1997 IEEE International Conference on Systems, Man and Cybernetics 1174–1179 (1997)
33. Virvou, M., Katsionis, G. Web services for an intelligent tutoring system that operates as a virtual reality game, Proceedings of the IEEE International Conference on Systems, Man and Cybernetics 1, 378–383 (2003)
34. Virvou, M., Tsiriga, V. Involving effectively teachers and students in the life cycle of an intelligent tutoring system, Educational Technology and Society 3(3), 511–521 (2000)
35. Vos, H.J. A Bayesian sequential procedure for determining the optimal number of interrogatory examples for concept-learning, Computers in Human Behavior 23(1), 609–627 (2007)
36. Walczak, S. Knowledge acquisition and knowledge representation with class: the object-oriented paradigm. Expert Systems with Applications, 15, 235–244 (1998)
37. Woolf, Cunningham. Multiple knowledge sources in intelligent teaching systems. IEEE Expert 2(2), 41–54 (1987)
38. Xie, X., Lam, K.-M. Elastic shape-texture matching for human face recognition, Pattern Recognition 41(1), 396–405 (2008)

2

Requirements Analysis and Design of an Affective Bi-Modal Intelligent Tutoring System: The Case of Keyboard and Microphone

Efthymios Alepis[1], Maria Virvou[1], and Katerina Kabassi[2]

[1] Department of Informatics, University of Piraeus, 80 Karaoli and Dimitriou St., 18534 Piraeus, Greece, talepis@unipi.gr, mvirvou@unipi.gr
[2] Department of Ecology and the Environment, Technological Educational Institute of the Ionian Islands, 2 Kalvou Sq., 29100 Zakynthos, Greece, kkabassi@teiion.gr

Summary. This chapter presents an affective bi-modal Intelligent Tutoring System (ITS) with emphasis on the early stages of its creation. Affective ITSs are expected to provide a more human-like interaction between students and educational software. The ITS is named Edu-Affe-Mikey and its tutoring domain is Medicine. The two modes of interaction presented in this chapter concern the keyboard and the microphone. Emotions of students are recognised by each modality separately and then, evidence from the two modalities is combined through a decision making theory. After emotion recognition has been performed Edu-Affe-Mikey adapts dynamically its tutoring behaviour to an appropriate emotion of an animated tutoring agent. In this respect an affective and adaptive interaction is achieved in the interactions of the student with the ITS, by performing both affect recognition and affect generation.

2.1 Introduction

Learning is a complex cognitive process and it is argued that how people feel may play an important role on their cognitive processes as well [9]. Indeed, as Coles [3] points out poor learning can produce negative emotions; negative emotions can impair learning; and positive emotions can contribute to learning achievement and vice versa. Therefore, a way of improving the learning process is recognising the users' emotions by observing them during their engagement with the educational software and then adapting its interaction to their emotional state.

Ahn and Picard [1] point out that affective biases from affective anticipatory rewards can be applied for improving the speed of learning, regulating the trade-of between exploration and exploitation in learning more efficiently. In Kim et al. [13] as well as in Burleson [2] it is suggested that pedagogical

E. Alepis et al.: *Requirements Analysis and Design of an Affective Bi-Modal Intelligent Tutoring System: The Case of Keyboard and Microphone*, Studies in Computational Intelligence (SCI) **104**, 9–24 (2008)
www.springerlink.com

agents with emotional interaction with learners should be used in learning environments. The findings of these studies imply that the emotional states of learning companions can be utilised to optimise students' motivation, learning and perseverance. For this purpose different approaches have been proposed in the literature. For example, Kapoor and Picard [12] propose a multi-sensor affect recognition system for children trying to solve an educational puzzle on the computer.

However, a problem that arises in such approaches in the way that these interaction modalities are combined in order to improve the accuracy and overall performance of emotion recognition in an Intelligent Tutoring System. In fact, the mathematical tools and theories that have been used for affect recognition can drive a classification of affect recognisers. Such a classification has been made in [14]. Liao and his colleagues, have classified affect recognisers into two groups on the basis of the mathematical tools that these recognisers have used: (1) the first group using traditional classification methods in pattern recognition, including rule-based systems, discriminate analysis, fuzzy rules, case-based and instance-based learning, linear and nonlinear regression, neural networks, Bayesian learning and other learning techniques. (2) The second group of approaches using Hidden Markov Models, Bayesian networks etc. Indeed, a recent piece of research uses the above approaches for the integration of audio-visual evidence [24]. Specifically, for person-dependent recognition, they apply the voting method to combine the frame-based classification results from both audio and visual channels. For person-independent test, they apply multistream hidden Markov models (HMM) to combine the information from multiple component streams. In contrast with these approaches we have used a decision making method in order to make decisions about which modalities should be taken to account while trying to recognise different emotional states. For this purpose two empirical studies were conducted and the results of these studies were used as criteria for the proposed decision making model.

A rather promising approach that has not attracted adequate attention despite of the potential benefits of its application is multi-criteria decision making theories. The main advantages of this approach derive from the fact that user-computer interaction is, by nature, multi-criteria-based. However, multi-criteria decision making requires several development steps for their application. More specifically, there is a need for experimental studies for selecting the criteria, estimating their weights of importance, test the models effectiveness, etc. Therefore, special emphasis should be given during requirements analysis and the design of an Intelligent Tutoring System (ITS). Indeed, as Virvou [23] point out, both students and tutors have to be involved in many phases of the software life-cycle in order to ensure that the software is really useful to students.

In view of the above, in this chapter, we discuss the requirements analysis and design of an ITS called Edu-Affe-Mikey. The proposed ITS is targeted to first-year medical students and its main characteristic is that it can adapt its interaction to each user's emotional state. For this purpose, the system uses a multi-criteria decision making method called Simple Additive Weighting (SAW) [8, 10] for combining two modes of interaction, namely keyboard and microphone. More specifically, SAW is used for evaluating different emotions, taking into account the input of the two different modes, and selects the one that seems more likely to have been felt by the user.

For requirements analysis and the effective application of the particular approach two different experimental studies have been conducted. The experimental studies involved real end users as well as human experts. In this way the application of the multi-criteria model in the system was more accurate as it was based on facts from real users' reasoning process. The main aim of the first study was to capture videos with real user's interaction with the system and as a result finding out how users express their emotions while interacting with educational software. The second empirical study involved human experts. These experts were asked to define the criteria that they usually use to perform emotion recognition of his/her students during the teaching course as well as their weights of importance.

The main body of this chapter is organised as follows: In Sect. 2.2 we present the multi-criteria decision making method called SAW. In Sect. 2.3 we present the first empirical study during requirements specification and analysis. In Sect. 2.4 we present the second empirical study for defining the criteria that are taken into account while performing emotion recognition. In Sect. 2.5 we give an overall description of the system and in Sect. 2.6 we give information of how the decision making method has been applied in the system for combining evidence from two different modes and select the user's emotion. Finally, in Sect. 2.7 we give the conclusions drawn by this work and discuss ongoing work.

2.2 SAW: A Simple Decision Making Method

A multi-criteria decision problem is a situation in which, having defined a set A of actions and a consistent family F of n criteria g_1, $g_2, \ldots,$ g_n $(n \geq 3)$ on A, one wishes to rank the actions of A from best to worst and determine a subset of actions considered to be the best with respect to F [22].

When the DM must compare two actions a and b, there are three cases that can describe the outcome of the comparison: the DM prefers a to b, the DM is indifferent between the two or the two actions are incompatible.

The traditional approach is to translate a decision problem into the optimisation of some function g defined on A. If $g(a) > g(b)$ then the DM prefers a to b, whereas if $g(a) = g(b)$ then the DM is indifferent between the two.

For the calculation of the function g many different decision making theories are introduced in the literacy. The Simple Additive Weighting (SAW) [8,10] method is among the best known and most widely used decision making method. SAW consists of two basic steps:

1. *Scale the values of the n criteria to make them comparable.* There are cases where the values of some criteria take their values in [0,1] whereas there are others that take their values in [0,1000]. Such values are not easily comparable. A solution to this problem is given by transforming the values of criteria in such a way that they are in the same interval. If the values of the criteria are already scaled up this step is omitted.
2. *Sum up the values of the n criteria for each alternative.* As soon as the weights and the values of the n criteria have been defined, the value of a multi-criteria function is calculated for each alternative as a linear combination of the values of the n criteria.

The SAW approach consists of translating a decision problem into the optimisation of some multi-criteria utility function U defined on A. The decision maker estimates the value of function $U(X_j)$ for every alternative X_j and selects the one with the highest value. The multi-criteria utility function U can be calculated in the SAW method as a linear combination of the values of the n criteria:

$$U(X_j) = \sum_{i=1}^{n} w_i x_{ij} \tag{2.1}$$

where X_j is one alternative and x_{ij} is the value of the i criterion for the X_j alternative.

2.3 Requirements Analysis for Affective Bi-Modal Interaction

Requirement specification and analysis in the affective bi-modal intelligent tutoring system resulted from an empirical study. The main aim of this study was to find out how users express their emotions through a bi-modal interface that combines voice recognition and input from keyboard. This empirical study involved 50 users (male and female), of the age range 17–19 and at the novice level of computer experience. The particular users were selected because such a profile describes the majority of first year medical students in a Greek university, which the educational application is targeted to. They are usually between the age of 17 and 19 and usually have only limited computing experience, since the background knowledge required for medical studies does not include advanced computer skills.

In the first phase of the empirical study these users were given questionnaires concerning their emotional reactions to several situations of computer use in terms of their actions using the keyboard and what they say. Participants were asked to determine what their possible reactions would be when they are at certain emotional states during their interaction. Our aim was to recognise the possible changes in the users' behaviour and then to associate these changes with emotional states like anger, happiness, boredom, etc.

After collecting and processing the information of the empirical study we came up with results that led to the design of the affective module of the educational application. For this purpose, some common positive and negative feelings were identified. These data were used for identifying the criteria that are taken into account when evaluating an emotion and a database was built for acquiring the weights of importance of these criteria.

The empirical study also revealed that the users would also appreciate if the system adapted its interaction to the users' emotional state. Therefore, the system could use the evidence of the emotional state of a user collected by a bi-modal interface in order to re-feed the system, adapt the agent's behaviour to the particular user interacting with the system and as a result make the system more accurate and friendly.

One conclusion concerning the combination of the two modes in terms of emotion recognition is that the two modes are complementary to each other to a high extent. In many cases the system can generate a hypothesis about the emotional state of the user with a higher degree of certainty if it takes into account evidence from the combination of the two modes rather than one mode. Happiness has positive effects and anger or boredom have negative effects that may be measured and processed properly in order to give information that is used during a human-computer affective interaction. For example, when the rate of typing backspace of a user increases, this may mean that the user makes more mistakes due to a negative feeling. However this hypothesis can be reinforced by evidence from speech if the user says something bad that expresses negative feelings.

2.3.1 Affect Perception Based on Speech

A small percentage of the participants say something with anger when they make a spelling mistake. However from the participants who do say something, 74% of them consider themselves having little or moderate (2–6 months of practice) computer knowledge. One very interesting result is that people seem to be more expressive when they have negative feelings (Fig. 2.1), than when they have positive feelings.

Another important conclusion coming up from the study is that when people say something either expressing happiness or anger it is highly supported from the changes of their voice. They may raise the tone of their voice or more probably they may change the pitch of their voice, Fig. 2.2.

Moreover it is interesting to notice that a very high percentage (85%) of young students who are also inexperienced with computers find the oral mode of interaction very useful.

2.3.2 Affect Perception Based on Keyboard Actions

While using the keyboard most of the participants agree that when they are nervous the possibility of making mistakes increases rapidly. This is also the

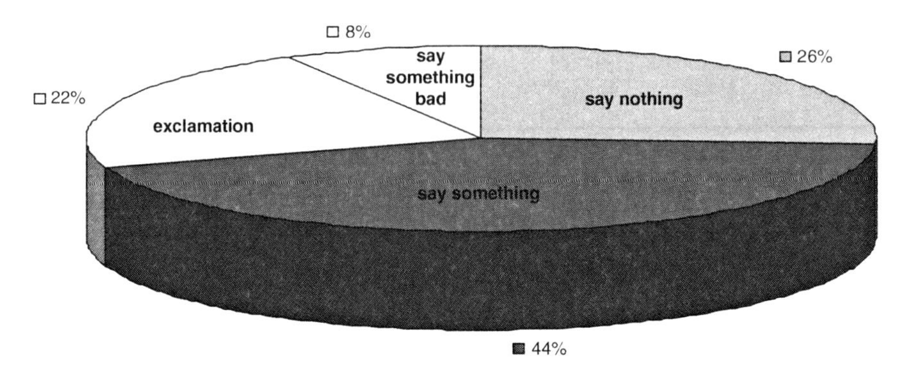

Fig. 2.1. Speech reactions in negative feelings

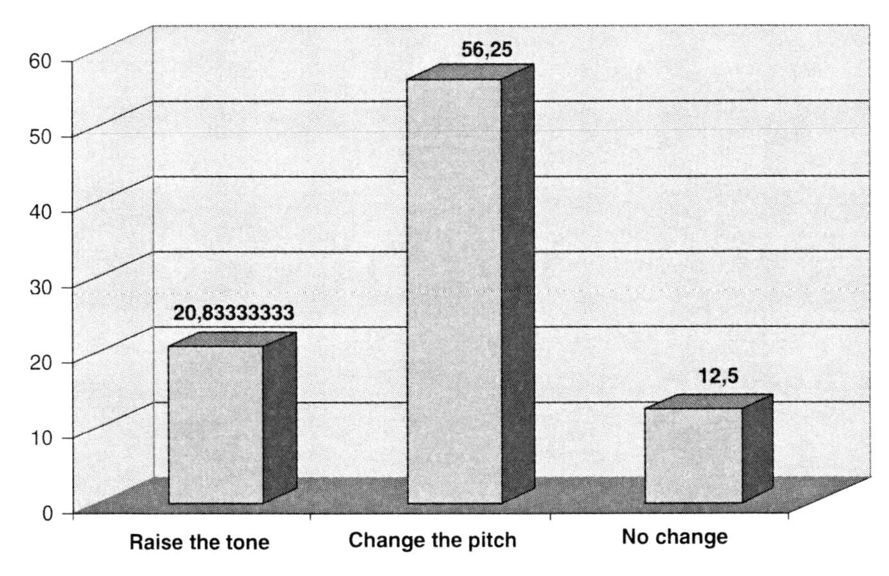

Fig. 2.2. Changes in voice when expressing a feeling (either positive or negative)

case when they have negative feelings. Mistakes in typing are followed by many backspace-key keyboard strokes and concurrent changes in the emotional state of the user in a percentage of 82%. Yet users under 20 years old seem to be more prone to making even more mistakes as a consequence of an initial mistake and lose their concentration while interacting with an application (67%). Students also admit that when they are angry the rate of mistakes increases, the rate of their typing becomes slower 62% (on the contrary, when they are happy they type faster 70%) and the keystrokes on the keyboard become harder (65%). Of course the pressure of the user's fingers on the keyboard is something that can not be measured without the appropriate hardware. Similar effects to the keyboard were reported for the neutral emotional state instead of anger.

2.4 Specification and Analysis of Multiple Criteria

Any decision making method requires prior to its application the specification of some criteria. Therefore, an empirical study was conducted in order to locate the criteria that human experts take into account while performing emotion recognition. The empirical study should involve a satisfactory number of human experts, who will act as the human decision makers and are reviewed about the criteria that they take into account when providing individualised advice. Therefore, in the experiment conducted for the application of the multi-criteria theory in the e-learning system, 16 human experts were selected in order to participate in the empirical study. All the human experts possessed a first and/or higher degree in Computer Science.

The participants of the empirical study were asked which input action from the keyboard and the microphone would help them find out what the emotions of the users were. From the input actions that appeared in the experiment, only those proposed by the majority of the human experts were selected. In particular considering the keyboard we have:

(a) User types normally
(b) User types quickly (speed higher than the usual speed of the particular user)
(c) User types slowly (speed lower than the usual speed of the particular user)
(d) User uses the backspace key often
(e) User hits unrelated keys on the keyboard
(f) User does not use the keyboard

Considering the users' basic input actions through the microphone we have seven cases:

(a) User speaks using strong language
(b) Users uses exclamations
(c) User speaks with a high voice volume (higher than the average recorded level)
(d) User speaks with a low voice volume (low than the average recorded level)
(e) User speaks in a normal voice volume
(f) User speaks words from a specific list of words showing an emotion
(g) User does not say anything.

2.5 Overview of the System

In this section, the overall functionality and emotion recognition features of our system, Edu-Affe-Mikey is described. The architecture of Edu-Affe-Mikey consists of the main educational application with the presentation of theory and tests, a programmable human-like animated agent, a monitoring user modelling component and a database.

While using the educational application from a desktop computer, students are being taught a particular medical course. The information is given in text form while at the same time the animated agent reads it out loud using a speech engine. The student can choose a specific part of the human body and all the available information is retrieved from the systems' database. In particular, the main application is installed either on a public computer where all students have access, or alternatively each student may have a copy on his/her own personal computer. An example of using the main application is illustrated in Fig. 2.3. The animated agent is present in these modes to make the interaction more human-like.

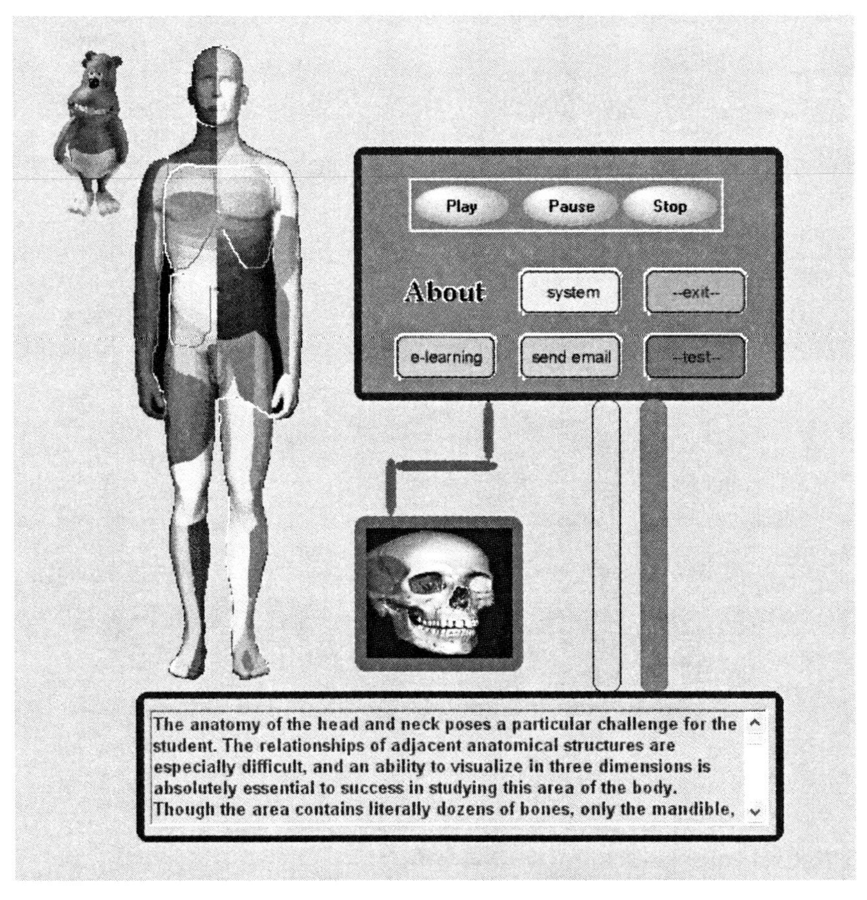

Fig. 2.3. A screen-shot of theory presentation in Edu-Affe-Mikey educational application

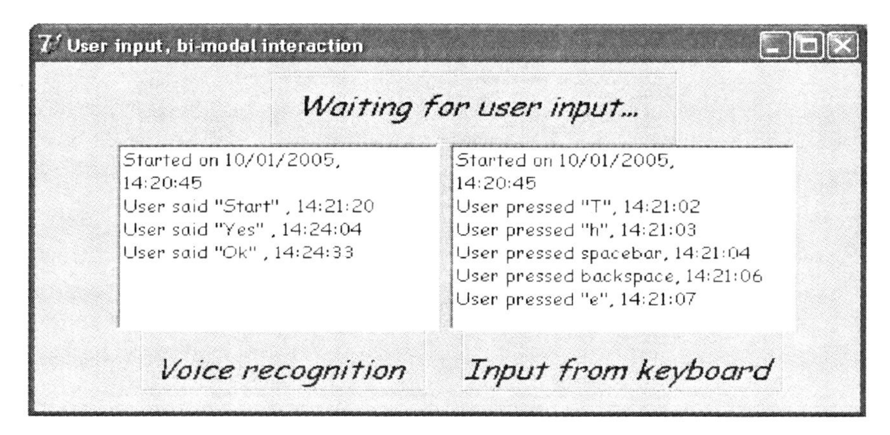

Fig. 2.4. Snapshot of operation of the user modelling component

While the users interact with the main educational application and for the needs of emotion recognition a monitoring component records the actions of users from the keyboard and the microphone. These actions are then processed in conjunction with the multi-criteria model and interpreted in terms of emotions. The basic function of the monitoring component is to capture all the data inserted by the user either orally or by using the keyboard and the mouse of the computer. The data is recorded to a database and the results are returned to the basic application the user interacts with. Figure 2.4 illustrates the 'monitoring' component that records the user's input from the microphone and the keyboard and the exact time of each event.

Instructors have also the ability to manipulate the agents' behaviour with regard to the agents' on screen movements and gestures, as well as speech attributes such as speed, volume and pitch. Instructors may programmatically interfere to the agent's behaviour and the agent's reactions regarding the agents' approval or disapproval of a user's specific actions. This adaptation aims at enhancing the effectiveness of the whole interaction. Therefore, the system is enriched with an agent capable to express emotions and, as a result, enforces the user's temper to interact with more noticeable evidence in his/her behaviour.

Figure 2.5 illustrates a form where an instructor may change speech attributes. Within this context the instructor may create and store for future use (Fig. 2.6) many kinds of voice tones such as happy tone, angry tone, whisper and many others depending on the need of a specific affective agent-user interaction. In some cases a user's actions may be rewarded with a positive message by the agent accompanied by a smile and a happy tone in the agent's voice, while in other cases a more austere behaviour may be desirable for educational needs. Figure 2.7 illustrates how an instructor may set possible actions for the agent in specific interactive situations while a user takes a

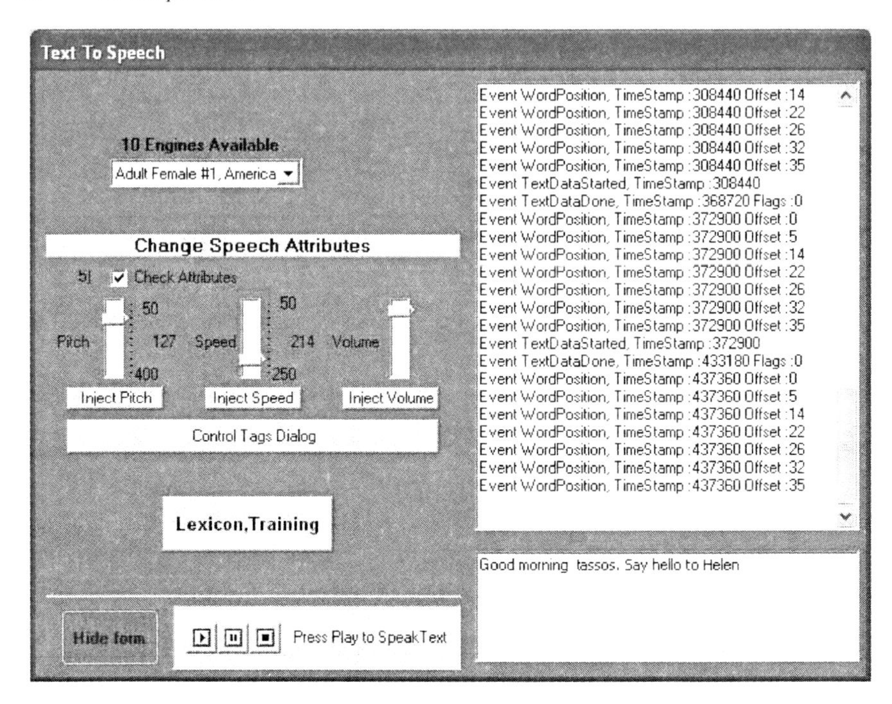

Fig. 2.5. Setting parameters for the voice of the tutoring character

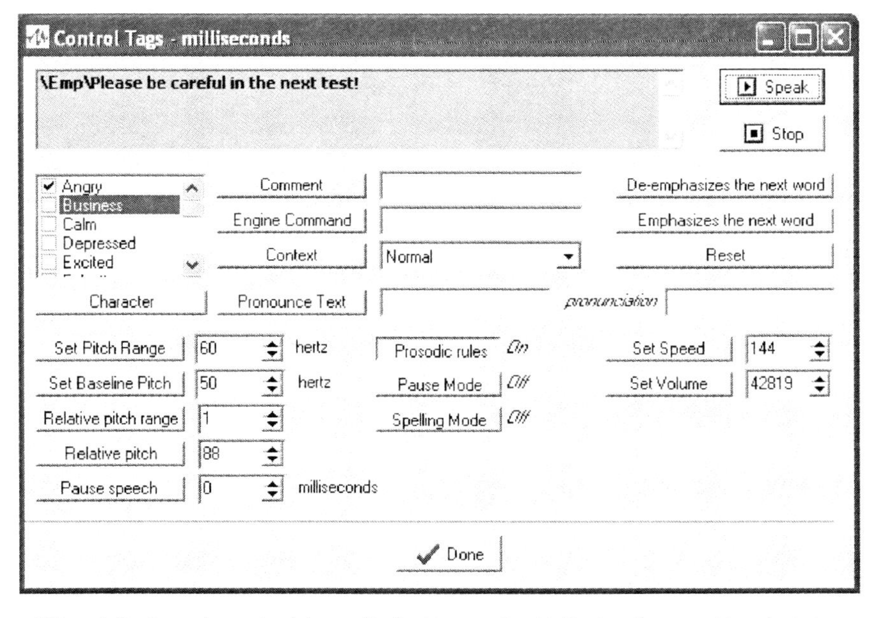

Fig. 2.6. Speech control tags. Default speech attributes for emotional states

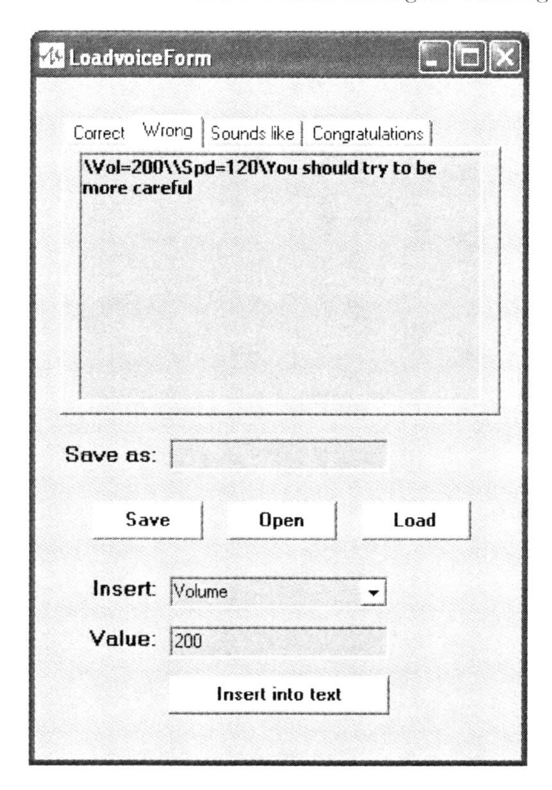

Fig. 2.7. Programming the behaviour of animated agents depending on particular students' actions

test. Instructors have also the potential of authoring their own tests for specific parts of the theory as to adapt the educational process to each individual student (Fig. 2.8). Finally Fig. 2.9 illustrates the interaction between a student and the animated agent while a student is taking a test. The system monitors all user actions and in cases where a student's action and a specified by an instructor event are correlated, the agent performs a pre-programmed behavior (visual and oral) that corresponds to that event.

2.6 Application of the Decision Making Method

For the evaluation of each alternative emotion the system uses as criteria the input actions that relate with the emotional states that may occur while a user interacts with an educational system. These input actions were identified by the human experts during the second experimental study and are considered as criteria for evaluating all different emotions and selecting the one that seems to be more prevailing. More specifically, the system uses SAW for a particular category of users. This particular category comprises of the young (under the age of 19) and novice users (in computer skills). The likelihood for

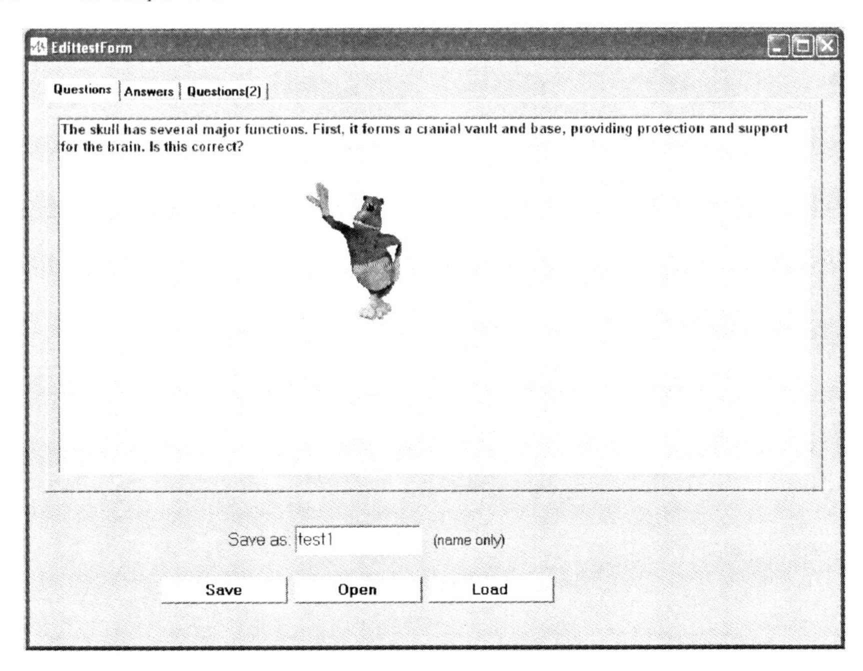

Fig. 2.8. Authoring tests for the educational application

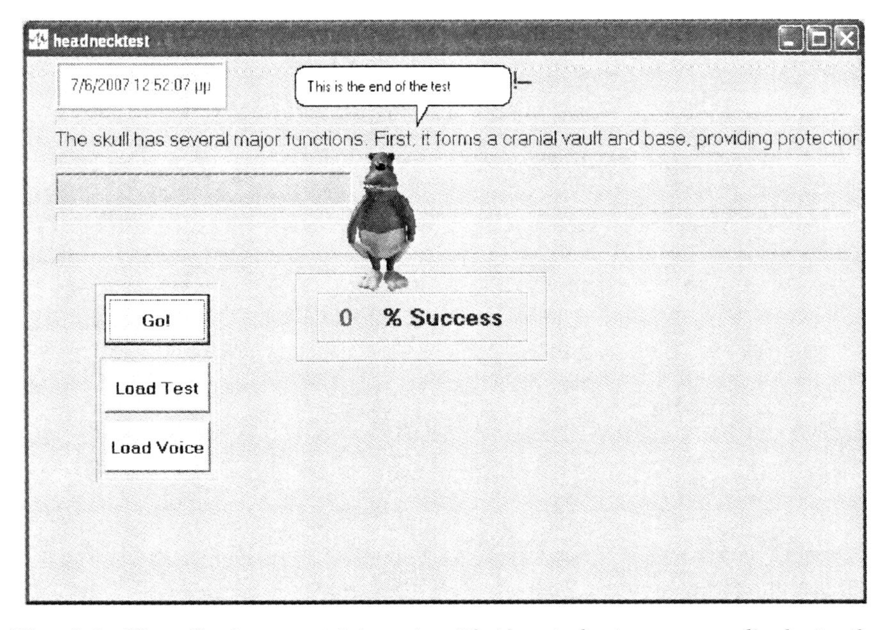

Fig. 2.9. The affective agent interacts with the student, correspondingly to the student's actions

a specific emotion (happiness, sadness, anger, surprise, neutral and disgust) to have occurred by a specific action is calculated using the formula below:

$$\frac{em_{1e_11} + em_{1e_12}}{2}$$

$$em_{1e_11} = w_{1e_1k1}k_1 + w_{1e_1k2}k_2 + w_{1e_1k3}k_3 + w_{1e_1k4}k_4 + w_{1e_1k5}k_5 + w_{1e_1k6}k_6 \quad (2.2)$$

$$em_{1e_12} = w_{1e_1m1}m_1 + w_{1e_1m2}m_2 + w_{1e_1m3}m_3 + w_{1e_1m4}m_4 + w_{1e_1m5}m_5 + w_{1e_1m6}m_6 + w_{1e_1m7}m_7 \quad (2.3)$$

em_{1e_11} is the probability that an emotion has occurred based on the keyboard actions and em_{1e_12} is the probability that refers to an emotional state using the users' input from the microphone. These probabilities result from the application of the decision making model of SAW and are presented in (2.2) and (2.3) respectively. em_{1e_11} and em_{1e_12} take their values in [0,1].

In (2.2) the k's from *k1* to *k6* refer to the six basic input actions that correspond to the keyboard. In (2.3) the m's from *m1* to *m7* refer to the seven basic input actions that correspond to the microphone. These variables are Boolean. In each moment the system takes data from the bi-modal interface and translates them in terms of keyboard and microphone actions. If an action has occurred the corresponding criterion takes the value 1, otherwise its value is set to 0. The w's represent the weights. These weights correspond to a specific emotion and to a specific input action and are acquired by the database.

In order to identify the emotion of the user interacting with the system, the mean of the values that have occurred using (2.2) and (2.3) for that emotion is estimated. The system compares the values from all the different emotions and determines whether an emotion is taking effect during the interaction. As an example we give the two formulae with their weights for the two modes of interaction that correspond to the emotion of happiness when a user (under the age of 19) gives the correct answer in a test of our educational application. In case of em_{1e_11} considering the keyboard we have:

$$em_{1e_11} = 0.4k_1 + 0.4k_2 + 0.1k_3 + 0.05k_4 + 0.05k_5 + 0k_6$$

In this formula, which corresponds to the emotion of happiness, we can observe that the higher weight values correspond to the normal and quickly way of typing. Slow typing, often use of the backspace key and use of unrelated keys are actions with lower values of weights. Absence of typing is unlikely to take place. Concerning the second mode (microphone) we have:

$$em_{1e_12} = 0.06m_1 + 0.18m_2 + 0.15m_3 + 0.02m_4 + 0.14m_5 + 0.3m_6 + 0.15m_7$$

In the second formula, which also corresponds to the emotion of happiness, we can see that the highest weight corresponds to *m6* which refers to the 'speaking of a word from a specific list of words showing an emotion' action. The empirical study gave us strong evidence for a specific list of words. In the case of words that express happiness, these words are more likely to occur in a situation where a novice young user gives a correct answer to the system. Quite high are also the weights for variables $m2$ and $m3$ that correspond to the use of exclamations by the user and to the raising of the user's voice volume. In our example the user may do something orally or by using the keyboard or by a combination of the two modes. The absence or presence of an action in both modes will give the Boolean values to the variables $k1 \ldots k6$ and $m1 \ldots m7$.

A possible situation where a user would use both the keyboard and the microphone could be the following: The specific user knows the correct answer and types in a speed higher than the normal speed of writing. The system confirms that the answer is correct and the user says a word like 'bravo' that is included in the specific list of the system for the emotion of happiness. The user also speaks in a higher voice volume. In that case the variables $k1$, $m3$ and $m6$ take the value 1 and all the others are zeroed. The above formulae then give us $em_{1_{e_1}1} = 0.4^*1 = 0.4$ and $em_{1_{e_1}2} = 0.15^*1 + 0.3^*1 = 0.45$.

In the same way the system then calculates the corresponding values for all the other emotions using other formulae. For each basic action in the educational application and for each emotion the corresponding formula have different weights deriving from the empirical study. In our example in the final comparison of the values for the six basic emotions the system will accept the emotion of happiness as the most probable to occur.

2.7 Conclusions and Ongoing Work

In this chapter we have described an affective educational application that recognises students' emotions based on their words and actions that are identified by the microphone and the keyboard, respectively. The system uses an innovative approach that combines evidence from the two modes of interaction using a multi-criteria decision making theory.

The main advantage of the proposed approach is that the whole process is based on experimental studies in which real users participate. Therefore, potential users of the software and human experts have participated in an empirical study. The analysis of the results of the empirical study gave evidence for the design of the affective module of the educational application. As the affective module uses a multi-criteria decision making theory for combining evidence from the two different modes, the results of the empirical study were used for selecting the criteria and estimating their weights of importance.

In future work we plan to improve our system by the incorporation of stereotypes concerning users of several ages, educational backgrounds and computer knowledge levels. Moreover, there is ongoing research work in progress that exploits a third mode of interaction, visual this time [21], to add information to the system's database and complement the inferences of the user modelling component about users' emotions. The third mode is going to be integrated to our system by adding cameras and also providing the appropriate software, as for a future work.

Acknowledgements

Support for this work was provided by the General Secretariat of Research and Technology, Greece, under the auspices of the PENED-2003 program. A part of this work is reported in the International Journal of Intelligent Support Technologies, IOS Press, Volume 1, Number 3, 2008.

References

1. Ahn, H., Picard, R.W. Affective-cognitive learning and decision making: A motivational reward framework for affective agents, Lecture Notes in Computer Science, Vol. 3784 LNCS (2005), pp. 866–873
2. Burleson, W. Affective learning companions: strategies for empathetic agents with real-time multimodal affective sensing to foster meta-cognitive and meta-affective approaches to learning, motivation, and perseverance. PhD. Thesis, 2006
3. Coles, Gerald. Literacy, emotions, and the brain. Reading Online, March 1999. http://www.readingonline.org/critical/coles.html
4. Damasio, A.R. Descartes' Error: Emotion, Reason and the Human Brain, Grosset/Putnam, New York (1994)
5. Davidson, R.J., Scherer, K.R., Goldsmith, H.H. Handbook of Affective Sciences, Oxford, New York (2003)
6. Davidson, R.J., Pizzagalli, D., Nitschke, J.B., Kalin, N.H. Parsing the subcomponents of emotion and disorders of emotion: perspectives from affective neuroscience, In Handbook of Affective Sciences, Davidson, R.J., Scherer, K.R., Goldsmith, H.H. (eds.) (2003)
7. Elfenbein, H.A., Ambady, N. When familiarity breeds accuracy. Cultural exposure and facial emotion recognition, Journal of Personality and Social Psychology, 85(2) (2003) 276–290
8. Fishburn, P.C. Additive utilities with incomplete product set: Applications to priorities and assignments, Operations Research 15 (1967) 537–542
9. Goleman, D. Emotional Intelligence, Bantam Books, New York (1995)
10. Hwang, C.L., Yoon, K. Multiple attribute decision making: Methods and applications, Lecture Notes in Economics and Mathematical Systems 186, Springer, Berlin, Heidelberg, New York (1981)
11. Isbister, K., Höök, K. Evaluating affective interactions: Innovative approaches and future direction, Special Issue of the International Journal of Human Computer Studies (2006)
12. Kapoor, A. Picard, R.W. Multimodal affect recognition in learning environments, Proceedings of the 13th annual ACM international conference on Multimedia, pp. 677–682 (2005)

13. Kim, Y., Baylor, A.L., Shen, E. Pedagogical agents as learning companions: The impact of agent emotion and gender, Journal of Computer Assisted Learning, 23(3) (2007) 220–234
14. Liao, W., Zhang, W., Zhu, Z., Ji, Q, Gray, W.D. Toward a decision-theoretic framework for affect recognition and user assistance, International Journal of Human-Computer Studies 64 (2006) 847–873
15. Moriyama, T., Ozawa, S. Measurement of human vocal emotion using fuzzy control, Systems and Computers in Japan 32(4) (2001)
16. Oviatt, S. User-modeling and evaluation of multimodal interfaces, Proceedings of the IEEE, Institute of Electrical and Electronics Engineers (2003) 1457–1468
17. Pantic, M., Rothkrantz, L.J.M. Toward an affect-sensitive multimodal human-computer interaction, Proceedings of the IEEE, Institute of Electrical and Electronics Engineers 91 (2003) 1370–1390
18. Picard, R.W. Affective computing: Challenges, International Journal of Human-Computer Studies 59(1–2) (2003) 55–64
19. Rich, E. Users are individuals: Individualizing user models, International Journal of Man-Machine Studies 18 (1983) 199–214
20. Stathopoulou, I.O., Tsihrintzis, G.A. Detection and expression classification system for face images (FADECS), IEEE Workshop on Signal Processing Systems, Athens, Greece (2005)
21. Stathopoulou, I.O., Tsihrintzis, G.A. Automated processing and classification of face images for human-computer interaction applications, Intelligent Interactive Systems in Knowledge-based Environments, 104 (2007)
22. Vincke, P. Multicriteria Decision-Aid. Wiley, New York (1992)
23. Virvou, M. Software engineering of environmental computer-assisted learning, Proceedings of the International Conference on Ecological Protection of the Planet Earth, 2 (2001) 927–934
24. Zeng, Z.H., Tu, J.L., Liu, M., Huang, T.S., Pianfetti, B., Roth, D., Levinson, S. Audio-visual affect recognition, IEEE Transactions on Multimedia 9(2) (2007) 424–428

3

Estimating the Development Cost for Intelligent Systems

Stamatia Bibi and Ioannis Stamelos

Department of Informatics, Aristotle University of Thessaloniki, Aristotle University Campus, Thessaloniki 54 124, Greece, sbibi@csd.auth.gr, stamelos@csd.auth.gr; http://sweng.csd.auth.gr

Summary. In this chapter we suggest several estimation techniques for the prediction of the functionality and productivity required to develop an Intelligent System application. The techniques considered are Analogy Based Estimation, Classification trees, Rule Induction and Bayesian Belief Networks. Estimation results of each technique are discussed and several conclusions are drawn regarding the methods, the platform and the languages used for the development of Intelligent Systems. The data set used in the analysis is the publicly available ISBSG data set.

3.1 Introduction

Software cost estimation is the process of predicting the amount of effort or the productivity required for the completion of a software artifact. Typically software cost estimation involves initially an assessment of the project attributes and then the application of a method for the generation of an estimate. This estimate is used for a number of purposes including budgeting, trade off and risk analysis, project planning and control, and investment analysis. The potential advantages that arise from this procedure can explain the vast amount of literature that exists on this area covering a wide range of methods and practices for software cost estimation.

With the growing importance of Intelligent Systems (IS) in various application domains [10,28,36], many practitioners see the measurement of intelligent applications [22] as a particularly interesting area of research. IS support the decision-making process in organizations and public sectors. As a consequence, enterprises and organizations adopt IS in order to discover hidden information and improve their performance in various areas such as public relationships, customer satisfaction and services innovation.

Successful completion of a software project requires appropriate software development and project management processes. For project managers, cost estimation is one of the crucial steps at the beginning of a new software project [27]. In the field of software cost estimation there is a trend to calibrate

S. Bibi and I. Stamelos: *Estimating the Development Cost for Intelligent Systems*, Studies in Computational Intelligence (SCI) **104**, 25–45 (2008)
www.springerlink.com

estimation models with the needs of particular applications, e.g. Web applications [32]. IS development differs substantially from traditional software development because in such projects there are additional constraints due to the different development approach, the cost of computation, the hardware demands and the different programming languages, platforms and algorithms used. Additionally it is known that cost estimation models are sensitive to the data used for their construction. Estimation models present increased accuracy when the projects used to predict future projects are developed in similar environment and under the same constraints with the projects under estimation. One method has been proposed for estimating the cost of IS development [26] based on COCOMO II model [7] that demands a wealth of data regarding the projects variables. This model is very analytical as it requires the values of 23 cost variables and then calculates a parametric cost equation. Considering that often in the initialization of a project there is no available knowledge regarding the factors discussed in [26] we suggest in this chapter the exploitation of knowledge of previously completed projects. A low cost method for productivity analysis is learning from past projects to predict future ones. In this study we extract knowledge from Isbsg data set [15] regarding two types of IS, namely Decisions Support Systems (DSS) [35] and Knowledge Based Systems (KBS) [30]. Using this knowledge one should be able to: (a) Identify a productivity interval for the development of IS projects, (b) Discover possible influence of project attribute values on productivity (the utilization of certain tools, languages, databases used may increase or decrease productivity).

This chapter provides an in depth analysis of the productivity required to complete an IS application. We apply several machine learning techniques, namely Analogy based estimation (ABE), Classification trees, Rule induction and Bayesian analysis in order to estimate size, productivity and extract useful knowledge regarding the development of IS projects. It should be mentioned that the applications studied were implemented with non AI languages. To our knowledge there is no publicly available data set with specialized data on the development of IS projects, such as language, mining algorithm, computational costs and database size. As a consequence we limit our analysis on a sample of IS projects coming from a large multi-organizational data set containing several other types of projects as a well. Due to the different type of projects that appear in the data set, the data considered in the study are general and they do not provide more specific knowledge involving issues, mentioned earlier concerning purely IS development.

3.2 Related Work

There are many studies in literature regarding the implementation of data mining techniques [9,12,22] for the development of IS projects. Various methods, techniques and algorithms for implementing IS have been a research issue for several studies [14, 39]. In these studies, certain methods for extracting knowledge from data bases have been analyzed, implemented and evaluated in terms of computational time, predictive accuracy, descriptive accuracy and misclassification costs. Comparative studies for different kind of development techniques and algorithms address issues regarding the mining efficiency of the methods but they do not deal with issues involving the time needed for the implementation of the applications, the tools and the productivity required for such projects when implemented with different kind of data mining techniques.

There are only few studies in literature dealing with software engineering issues in the development of IS. In [13] a software engineering environment for developing IS is suggested, that enables the cooperation of various technologies for that purpose. The reusability of database components utilized in the development of IS is discussed in [38]. Organizing IS development process with Crisp DM methodology is suggested in [9] that provides a framework that contains the corresponding phases of a project, their respective tasks, and relationships between these tasks. These studies propose methods for simplifying the development of IS projects. Another approach to simplify the development of IS projects is to effectively schedule the whole software engineering process of IS systems. One part of scheduling involves the estimation of costs. An estimation of the time needed to complete an IS project and of the way the selection of software development language, platform or methodology affects this time could provide useful knowledge especially in the initialization of an IS project. One method has been proposed for estimating the cost of IS development [26] based on COCOMO II model [7] that utilizes data regarding the projects variables. This model is very analytical as it requires the values of 23 cost variables, such as number of tables and tuples, type of data sources, type and number of models extracted, and then calculates a parametric cost equation. To our knowledge no other study is found in literature regarding specifically IS development costs.

On the other hand there is a vast literature in the area of software cost estimation (SCE). An extensive review of software cost estimation methods can be found in [6]. Expert judgment [19–21, 29], Model based techniques [1, 7, 16, 17, 33] and Learning oriented techniques [3, 4, 24, 25] are the main categories of software cost estimation methods. The wide application of the techniques in several development environments for the estimation of software costs has indicated that the combination of methods provides improved results. Cost estimation techniques can present several advantages and disadvantages in terms of accuracy, comprehensibility, causality, applicability and risk management. The combination of methods can provide a more complete estimation framework than one method alone. In this study taking into consideration the findings of [23] we suggest the combination of several learning oriented methods in order to improve the estimation and descriptive efficiency of predictive models for an IS project.

Learning oriented techniques, are recently applied in SCE and use data from past historical projects to estimate and control the new ones. Learning oriented techniques are very useful when a company stores its own data and wants to utilize this knowledge in order to guide and control the development of future projects. Literature has indicated several advantages and disadvantages of these methods when applied alone.

ABE is a widely used method to identify historical projects, with attributes similar to the project under estimation, that will provide an estimation [2]. ABE is a very accurate method when the historical data set includes similar projects to the one under estimation but can be very inaccurate in other situations.

Neural networks [25] can be very accurate but difficult to apply and interpret, while rule induction and decision trees [24] are easy to understand but of lower accuracy. Association rules as a descriptive modeling method [4, 5] can address causality, are easily interpretable but they cannot always provide an estimate. Bayesian Networks [3, 34] identify the underlying relationships among all project attributes, can be easily combined with expert judgment but usually they demand large data sets in terms of accuracy.

In this study we experiment with several learning methods regarding the SCE of IS in order to provide a complete estimation framework. In particular we apply:

- Rule induction and Classification And Regression Trees (CART): to estimate size and productivity with probability values that assess uncertainty.
- ABE: to identify important attributes and distance metrics for productivity and size estimation.
- Association rules: to explore patterns that associate various project attributes.
- Bayesian Belief Networks (BBN): to identify the dependencies among the project attributes.

3.3 Estimation Methods

In this section we present analytically the methods used to analyze and model data regarding the implementation of IS systems.

3.3.1 Classification and Regression Trees

CART is a widely used statistical procedure for producing classification and regression models with a tree-based structure in predictive modeling [8]. The CART model consists of an hierarchy of univariate binary decisions. The algorithm used operates by choosing the best variable for splitting data into two groups at the root node. It can use any one of several different splitting criteria, all producing the effect of partitioning the data at an internal node into two disjoint subsets in such way that the class labels are as homogeneous as possible. This splitting procedure is then applied recursively to the data in each of the child nodes. A greedy local search method to identify good candidate tree structures is used. Finally, a large tree is produced and specific

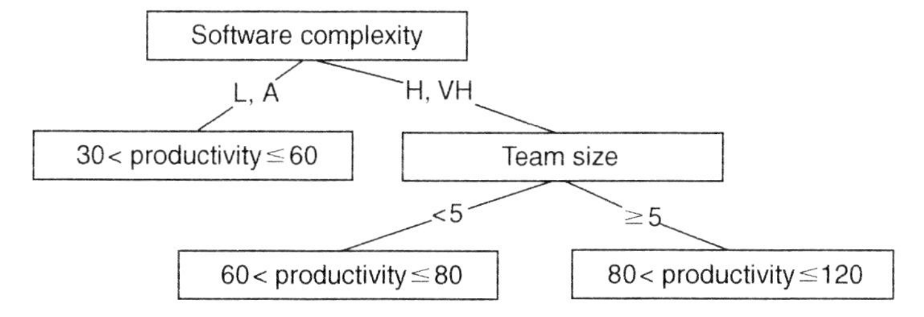

Fig. 3.1. A CART for software productivity estimation

branches of this tree are pruned according to the stopping criteria, so as to avoid overfitting of the data and over-specialization of the model.

CART are able to classify not only all the projects in the training data set, but unknown projects from a wider group of projects of which the training projects is presumed to provide a representative example. In our study a serial algorithm is applied. The decision tree classifier consists of two phases: a *growth* phase and a *prune* phase. In the growth phase, the tree is built by recursively partitioning the data until each partition is either 'pure' or sufficiently small. The form of the split used to partition the data depends on the type of the attribute used in the split. Only binary splits are considered. Once the tree is fully grown, it is 'pruned' in the second phase to generalize the tree by removing dependence on statistical noise. Pruning a branch of a tree consists of deleting all descendants of the branch except from the root node. The pruning algorithm used is based on the Minimum Description Length principle [31].

A simple CART for software productivity estimation is presented in Fig. 3.1. In order to estimate the productivity category of a new project, starting from the root node, the right branch is selected according to the project's attributes at each level, moving down until a terminal node is reached. In particular the tree of Fig. 3.1 can be interpreted as following:

If software complexity is average or low productivity value is estimated to be greater than 30 fp h^{-1} and equal or less than 60 fp h^{-1}. If software complexity is high or very high and the team size is less than five people productivity value is estimated from 60 to 80 fp h^{-1}. Otherwise if the team size includes five or more people productivity will be between 80 and 120 fp h^{-1}.

We use classification trees for the analysis because they handle variables with nominal and ordinal scales, they provide results easy to interpret, and they can handle data sets with few cases.

3.3.2 Association Rules

Association rules [14] are among the most popular representations of local pattern recognition. They are a form of descriptive modeling and have as a target to describe the data and their underlying relationships with a set of rules that jointly define the target variables [39]. Their target is to find frequent combinations of attribute values that lay in databases. An association rule is a simple probabilistic statement about the co-occurrence of certain events in a database.

Each rule consists of two parts. The left part is the Rule Body (antecedent) and is the necessary condition in order to validate the right part, Rule Head (consequent). Each rule states that if the rule body is true then the rule head is also true with probability p. It is obvious that the A.R are Boolean propositions with true or false values. In the rule head, any Boolean expression can be used, but usually conjunction is preferred for simplicity purposes.

Given a set of observations over attributes $\mathrm{A}_1, \mathrm{A}_2, \ldots, \mathrm{A}_\mathrm{n}$ in a data set D a simple association rule has the following form:

$$(\mathrm{A}_1 = \mathrm{X} \wedge \mathrm{A}_2 = \mathrm{Y}) \Rightarrow \mathrm{A}_3 = \mathrm{Z}$$

$$\text{confidence} = p(A_3 = Z | A_1 = X, A_2 = Y), \ \text{support} = \text{freq}\ (X \cup Y \cup Z, D)$$

This rule is interpreted as following: when the attribute A_1 has the value X and attribute A_2 has the value Y then there is a probability p (confidence) that attribute A_3 has the value Z. For this rule two major statistics are computed, confidence and support values. Confidence is the probability p defined as the percentage of the records containing X, Y and Z with regard to the overall number of records containing X and Y only. The other statistic is support value which is a measure that expresses the frequency of the rule and is the ratio between the number of records that present X, Y and Z to the total number of records in the dataset.

AR mining is a two stage process. The first stage involves the identification of all frequent set of attributes contained in the given data set. A set of attributes is frequent if its associated support exceeds a certain support threshold defined by the user. The second stage is generating all pertinent ARs from these itemsets. An AR is pertinent if its associated confidence exceeds a certain confidence threshold specified by the user.

A simple example of an Association Rule coming from software data sets is presented in Table 3.1.

The association rule of Table 3.1 states that 80% of the cases that involve the development of a *DSS* project *Inhouse* also use a *Methodology*. If the data set used contains 39 instances then 12 out of 39 projects (support $= 12/39 = 30.8\%$) of the data set satisfy the particular rule and 12 out of 14 projects (confidence $= 12/14 = 80\%$) that involve the development of a *DSS* project *Inhouse* use a *Methodology*.

Table 3.1. Association rule describing whether methodology is used or not

Support	Confidence	Rule Body	Rule Head
30.8	80.0	DSS + DevelopedInhouse	Methodology_Yes

AR was selected because it is a method for descriptive modeling that identifies specific relationships often ignored by predictive models. Frequent and pertinent relationships among the project attributes and the development productivity are discovered. The knowledge extracted from the data set is in the form of *if-then* rules that are easily understood, giving the chance to experts to analyze, validate and combine known facts about the domain.

3.3.3 Bayesian Belief Networks

Bayesian Belief Networks are Directed Acyclic Graphs (DAGs), which are causal networks that consist of a set of nodes and a set of directed links between them, in a way that they do not form a cycle [18]. Each node represents a random variable that can take discrete or continuous finite, mutually exclusive values according to a probability distribution, which can be different for each node. Each link expresses probabilistic cause-effect relations among the linked variables and is depicted by an arc starting from the influencing variable (parent node) and terminating on the influenced variable (child node). The presence of links in the graph may represent the existence of direct dependency relationships between the linked variables (that some times may be interpreted as causal influence or temporal precedence). The absence of some links means the existence of certain conditional independency relationships between the variables.

The strength of the dependencies is measured by means of numerical parameters such as conditional probabilities. Formally, the relation between the two nodes is based on Bayes' Rule [11]:

$$P(A|B) = \frac{P(B|A)P(A)}{P(B)}. \tag{3.1}$$

For each node A with parents B1, B2, ..., Bn there is attached an NxM Node Probability Table (NPT), where N is the number of node states and M is the product of its cause-nodes states. In this table, each column represents a conditional probability distribution and its values sum up to 1.

A simple BBN estimating software effort is the one presented in Fig. 3.2. Attached to the node of effort there is a node probability table that provides possible values of the effort needed to develop the intelligent system, based on the combination of values that the programming language and the development platform nodes take.

In this study the extraction of BBN is achieved with an award winning algorithm [11]. BBN were selected because of their ability to represent all the dependencies among the variables participating in the study.

3.3.4 Rule Induction

RI, as CART method, is a particular aspect of inductive learning. Inductive learning is then the process of acquiring general concepts from specific

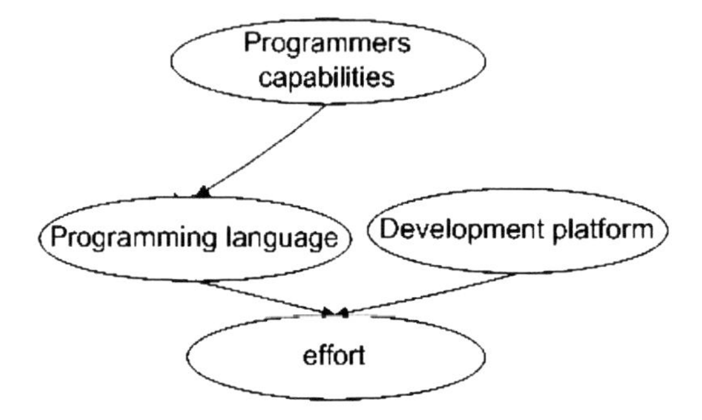

Fig. 3.2. A BBN for software effort estimation

Table 3.2. Rule induction for software productivity estimation

If language used = java and development type = enhancement then $40 < \text{productivity} \leq 60$ total no of projects = 10 wrong estimates = 2

examples. By analyzing many examples, it may be possible to derive a general concept that defines the production conditions.

Rule induction is an alternative approach to CART that takes each class separately, and try to cover all examples in that class, at the same time excluding examples not in the class. This is a so called, *covering approach*, because at each stage a rule is determined that covers some of the examples. Covering algorithms operate by adding tests to the rule that is under construction, always trying to create a rule with maximum accuracy. Whereas CART algorithm chooses an attribute to maximize the separation between the classes (using information gain criterion), the covering algorithm chooses an attribute-value pair to maximize the probability of the desired classification.

In this chapter, we apply the PART algorithm that is based on extracting partial decision trees. The method also handles missing values by assigning an instance with missing value to each of the tree branches with a weight proportional to the number of training instances descending that branch, normalized by the total number of training instances with known values. In this study for the induction of classification trees and rules we utilize the Weka machine learning library [37].

A simple rule coming from the domain of software cost estimation will have as Rule Body certain software project attribute values and as a Rule Head a productivity (or cost, or effort) value. A simple example of a rule is presented in Table 3.2.

This rule is interpreted as following: If the language that will be used for the development of new project is java and the development type of the project is enhancement then there is (10-2)/10 = 80% (confidence value) probability that the productivity value of the project will be between 40 and 60 lines of code per hour.

One advantage of inductive learning over other machine learning methods is that the rules are transparent and therefore can be read and understood. Proponents of RI argue that this helps the estimator understand the predictions made by systems of this type.

3.3.5 Analogy Based Estimation

ABE is essentially a form of case based reasoning. The main aspect of the method is the utilization of historical information from completed projects with known size, effort or productivity. The most appropriate attributes are selected according to which the new project is compared with the old ones in the historical dataset. The attribute values are standardized (between 0 and 1) so that they have the same degree of influence and the method is immune to the choice of units.

The next step is to calculate how much the new project differs from the other projects in the available database. This can be done by using a 'distance' metric between two projects, based on the values of the selected attributes for these projects. The most known such distance metric is the *Euclidean* or *straight-line distance* which has a straightforward geometrical meaning as the distance of two points in the k-dimensional Euclidean space:

$$d_{new,i} = \left\{ \sum_{j=1}^{k} (Y_j - X_{ij})^2 \right\}^{1/2}, \quad i = 1, 2, ..., n \tag{3.2}$$

Other possible distance metrics are the Minkowski distance, the Canberra distance, the Czekanowski coefficient and the Chebychev or 'Maximum' distance (see [2] for definitions).

In conclusion the estimation of the productivity using analogies is based on the completed projects that are similar to the new one. The user of the method has to calculate the distances of the new project from all the database projects and identify few 'neighbour' projects, i.e. those with relatively small distance value. The estimation of the productivity/effort is eventually obtained by some combination of the productivities/efforts of the neighbor projects. Typically, the statistic used is the mean (weighted or simple) or the median of these effort values. The calibration of the analogy-based method requires the detection of the best configuration of the available method options. The options that may be adjusted are:

(a) The distance metric by which the projects of the database will be sorted according to their similarity to the one under estimation (e.g. Euclidean distance, Manhattan distance)
(b) The number of closest projects (analogies)
(c) The set of attributes for judging analogy

Fig. 3.3. Best calculation parameters for effort estimation

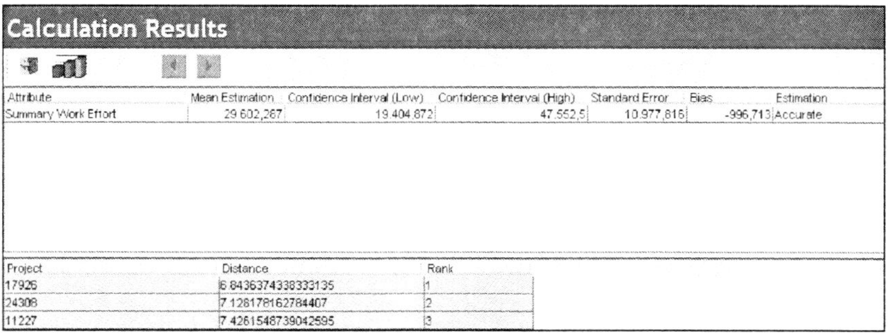

Calculation Results

Attribute	Mean Estimation	Confidence Interval (Low)	Confidence Interval (High)	Standard Error	Bias	Estimation
Summary Work Effort	29.602,267	19.404,872	47.552,5	10.977,816	-996,713	Accurate

Project	Distance	Rank
17926	6.8436374338333135	1
24308	7.128178162784407	2
11227	7.4261548739042595	3

Fig. 3.4. Estimation of the effort value of a new project based on the best parameters of Fig. 3.3

In this study we apply and calibrate ABE for size and productivity prediction with the help of Brace tool [2]. A simple example that demonstrates the calibration of the method and the selection of best parameters is presented in Fig. 3.3. In this example the most appropriate distance metric to be used for the calculation of distances among similar projects for the particular data set is Canberra distance, the size adjusted median of the predicted interval is selected as a point estimate and the number of analogies is set to three. The Mean Magnitude Relative Error (MMRE) is 26.9% and the prediction within 25% of actual value is 56%. The size adjusting attribute is function points, a choice made by the user.

The results of the application of ABE in the estimation are presented in Fig. 3.4. Similar projects from the training set to the project under estimation are presented in descending order. The effort values of the first three similar projects are used for the calculation of an effort interval of the new project. Apart from a low and a high effort value, a point estimate is calculated along with a bias value.

3.4 Data Description and Preparation

The data set used for this analysis is the widely known ISBSG7 data set, a publicly available multi-organizational data set. ISBSG is a repository maintained by the International Software Benchmarking Standards Group to help developers with project estimation and benchmarking. ISBSG data repository release 7 [15] contains 1,239 projects that cover the software development industry from 1989 to 2001. The data set contains over 50 fields involving the projects origin, age, context, the type of the product and the project and the development environment, the methods and tools utilised. We selected the particular data set because it contained 39 projects involving the development of Decision Support Systems and Knowledge Based Systems that could form a sample data set for our analysis. All variables were taken into account in the study regardless of the missing values observed. The variables are presented in Tables 3.3 and 3.4.

Table 3.3. Variables in ISBSG data set

Attribute	Values
DT: Development type	Enhancement, NewDevelopment, Re-development
LT: Language type	3GL, 4GL
PPL: Programming language	4GL, ACCESS, C, COBOL, CSP, JAVA, NATURAL, OTHER 4GL, VISUAL BASIC
OT: Organization type	Banking, Communication, Community Services, Electricity & Gas & Water, Insurance, Manufacturing, Professional Services, Public Administration, Computers
AT: Application type	DSS, Knowledge based
UCT: Use of case tools	Yes, No
PC: Package customization	Yes, No
BAT: Business area type	Banking, Engineering, Financial, Insurance, Manufacturing, Personnel, Sales & Marketing, Social Services, Hardware/Software
DBMS: Database system	ADABAS, ACCESS, IMS, SQL, ORACLE, RDB, M204
Used Methodology: whether methodology was used to build the software	Yes, No
DP: Development platform	MF, MR, PC
HMA: How methodology acquired	Developed/purchased, Developed Inhouse, Purchased

Table 3.4. Descriptive statistics for the attributes of IS systems

Variable	Min	Max	Mean	St. deviation
Function points	18	1,474	413.846	403.827
Effort	220	16,000	3,093.154	3,874.975
Productivity	0.024	1.147	0.28	0.306
Max team size	0	18	4.235	4.562
Date	1993	2000	1995.703	1.869

Table 3.5. Descriptive statistics for the attributes of MIS systems

Variable	Min	Max	Mean	St. deviation
Function points	25	17,518	557.276	1,097.789
Effort	17	99,088	6,412.665	11,107.226
Productivity	0.01	5.35	0.204	0.402
Max team size	0	468	8.184	36.071
Date	1989	2001	1996.201	2.871

In the data set there are two variables indicative of the projects cost, i.e. *size* measured in function points and *effort* measured in hours. Function Points (FP) is a standard metric for the relative size and complexity of a software system, originally developed by Alan Albrecht of IBM [1]. The size is determined by identifying the components of the system as seen by the end-user: the inputs, outputs, inquiries, interfaces to other systems, and logical internal files. The components are classified as simple, average, or complex. All of these attributes are then assigned a score by the analyst and the total is expressed in Unadjusted FPs (UFPs). Complexity factors described by 14 general systems characteristics, such as reusability, performance, and complexity of processing can be used to weight the UFP. Factors are also weighted on a scale of 0–5. The result of these computations is a number that correlates to system functional size. We decided to estimate the productivity value of the projects as well defined as the ratio between the function points of a project and the effort measured in hours.

In the same data set there are data regarding the implementation of other application type projects. It is interesting to compare such data with those of the IS data set. In Table 3.5, descriptive statistics regarding the function points, effort, productivity, max team size and date values of Management Information Systems are presented. Maximum values of function points between the two application types are quite different but the min and mean values are close for both types. Regarding the effort we can conclude that in the particular data set a simple MIS project requires less effort than an IS application but in total the mean effort required for a MIS project is more than the mean effort required for an IS application. Mean productivity values for both application types are quite similar although MIS projects require larger development teams, almost double in size, than IS projects.

Table 3.6. Productivity and function points intervals considered in the study

Variable	Low	Average	High	Very high
Productivity	0.024–0.09	0.09–0.168	0.168–0.6075	0.6075–1.147
Function points	18–157	157–254.5	254.5–916	916–1,474

Finally, as expected MIS application had earlier implementation dates than IS applications. We must mention that this comparison is limited to the specific information included in ISBSG data set. Also the observations regarding IS projects are only 39 and the ones regarding MIS projects are 352.

In order to apply the estimation methods we had to discretize [5] the field of productivity and assign productivity values into intervals. Discretization of productivity raised two issues: the number of intervals and the discretization method. Due to the few projects contained in the data set we intuitively decided to define four intervals of productivity each one related to approximately equal number of projects. The productivity (measured in function points per hour), and function point intervals are presented in Table 3.6.

Especially for Bayesian analysis data preparation was slightly different as the method cannot handle missing values. We had to dismiss records with missing values in several variables so as to leave as many projects and variables as possible in the analysis. The final set of projects was 20, the number of productivity intervals 3 0.02–0.1, 0.1–0.4, 0.4–1.15 and the variables that participated are the ones in Tables 3.4 and 3.5 apart from *dbms, hma* and *bat* that presented many missing values. Also for variables like organization type and language that presented many different values we grouped together values that had the same effect on productivity.

3.5 Results and Discussion Regarding Functional Size

In this section the results regarding the estimation of the *size* of IS applications will be presented in the form of classification tree and rules. The classification tree suggested is presented in Fig. 3.5. It seems that the *business area type* in which the system will be applied and the *data base* used can affect the *size* of the application. Applications implemented for financial and engineering business area types are relatively large in size. For the rest of the applications, the ones that used Adabas, Access, Oracle and Relational Data Bases as a data base management system presented average size while the rest could be considered as small applications. The classification and regression tree of figure can classify correctly 53.85% of the data set.

PART decision rules regarding the estimation of size are presented in Table 3.7. The rules comply with the CART of Fig. 3.5 and additionally provide two hints: Applications developed in *Access* with the help of *Case tools* present average size. *Decision support systems* also are of average size. The

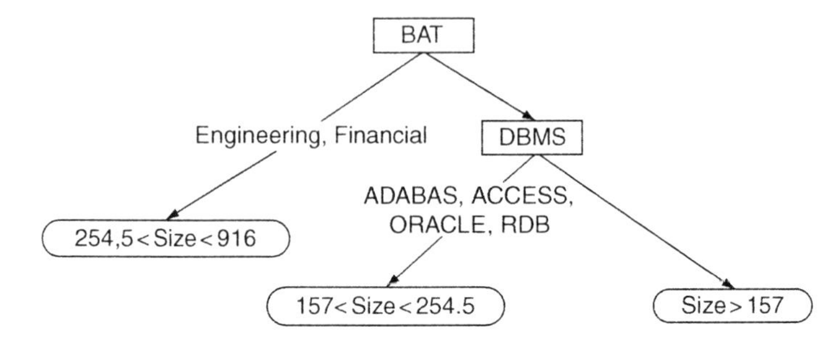

Fig. 3.5. CART for the estimation of functionality of an IS

Table 3.7. Rule set for the estimation of function points

Rules
BAT = Financial: 254.5 < size < 916 (6.78/2.26)
Dbms = Access AND CaseTool = Yes: 157 < size < 254.5 (4.33/1.41)
BAT = Engineering: 254.5 < size < 916 (3.52/1.23)
AT = DSS: > 157 < 254.5 (14.33/8.12)
else: size < 157 (10.04/5.42)

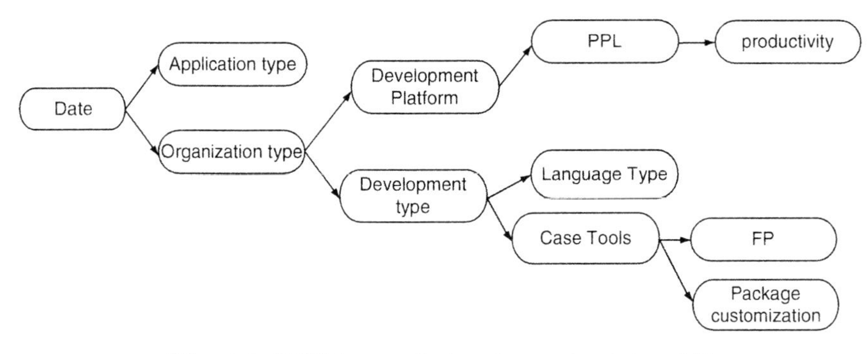

Fig. 3.6. BBN for predicting the productivity of IS

rule set of Table 3.7 is also able to classify correctly 53.85% of the projects into a size interval.

ABE pointed out as the best configuration to use the Euclidean distance metric with one analogy considering as best point estimate the mean value of the size interval. The best attribute subset for providing an estimation is considered the *development type*, the *development platform*, the *language type* the *data base* used and the *use of case tools.*

Bayesian analysis presented in Fig. 3.6 indicated the *use of case tools* as the only variable that directly affects software size. Also the number of function points is conditionally independent from *organization type* and *development type* when there is knowledge regarding the use of case tools. Since the values of case tools (no, yes) are uniformly distributed the only conclusion we can

Table 3.8. Node probability table for the estimation of function points

Use of case tools size	No	Yes
Size < 225	0.244	0.445
225–800	0.378	0.311
Size > 800	0.378	0.244

reach regarding the number of function points is that when there is absence of case tools the functionality that has to be implemented is increased which is a strange finding. The use of case tools enables the implementation of the same projects with less functionality implemented by the developers, a fact that reduces complexity and therefore the possibility of fault existence. The node probability table for the estimation of the number of function points is presented in Table 3.8. The last column of Table 3.8 can be interpreted as following: The functional size of an IS project when case tools are used to aid the development is 44.5% possible to be less than 225 fp, 31.1% is likely to be between 225 and 800 fp and there is only 24.4% possibility to be more than 800 fp.

3.6 Results and Discussion Regarding Productivity

This section contains the details of a formal analysis of the data set, comprising a classification tree analysis, rule induction and Bayesian analysis.

The tree extracted from the data is presented in Fig. 3.7. In the tree, the variables *Business Area Type, Function Points* and *Primary Programming Language* are able to classify correctly 69.23% of the projects. Low productivity values are observed in *Banking, Insurance, Personnel* and *Social Services* applications. High productivity value is presented in large applications developed mainly with *Cobol, Access, C* and other not mentioned *4GL* languages. The intermediate classes of productivity are obtained in relatively small application (<800 fp) or in applications developed with *Natural, Java, SQL* and *Csp*.

Table 3.9 presents the rule extracted with PART algorithm. The first number in brackets is the total number of projects that comply with the rule body while the second number is the number of the projects being misclassified by the rule in the subset of the examples the rule is applied.

The rules are applied in the order discovered and are able to classify correctly 71.8% of the projects participating in the study. Important attributes that appear often in the rules are *Organization and Application type. Function points* also seam to be a critical estimation variable. Low productivity values are observed in the development of projects designated to *Insurance* and *Public administration* organizations. Projects designed for *Energy* organizations are mainly assigned to high productivity values while projects applied

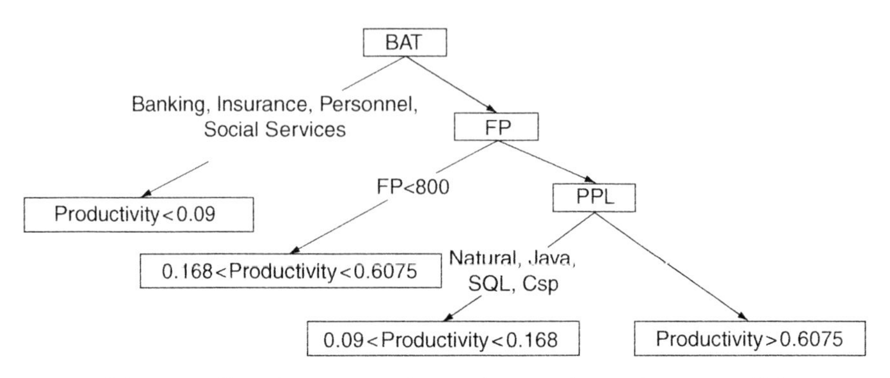

Fig. 3.7. Classification tree for predicting the productivity of IS

Table 3.9. List of rules for predicting the productivity of IS

Rules
OT = Communication AND FP $\leq$ 937: 0.168 < productivity < 0.6075 (10.0/4.33)
OT = Insurance: productivity < 0.09 (7.52/3.52)
AT = DSS AND OT = PublicAdministration: productivity < 0.09 (5.44/1.44)
OT = Energy: productivity > 0.6075 (3.21/0.21)
AT = DSS AND FP <= 722: 0.168 < productivity < 0.6075 (6.84/2.28)
AT = DSS: 0.09 < productivity < 0.168 (3.0/0)
else: productivity > 0.6075 (3.0/0)

Table 3.10. Node probability table for the estimation of the productivity of IS

ppl productivity	Access	Other 4GL, cobol, natural	sql,c,csp,java
Productivity < 0.1	0.185185	0.547619	0.138889
0.1–0.4	0.185185	0.261905	0.638889
Productivity > 0.4	0.62963	0.190476	0.222222

to *Communication* organizations tend to fall in the intermediate classes of productivity.

The results observed from the Bayesian analysis, presented in Fig. 3.6 are also very interesting. A strange absence of dependency in the BBN of Fig. 3.6 is the one between FP and productivity. This may be explained due to the small training set, the few intervals defined in each variable and the multi organizational data set that prevented the homogeneity of the data. Interesting dependency is the one between *programming language* used and *productivity* value. *Language type* can classify correctly 80% of the projects into a productivity interval with the help of Table 3.10.

Interesting assumptions regarding the Bayesian analysis can be summarized as following:

- Applications developed in *mainframe platforms* had mostly as destination *public administration organizations*, were programmed mainly with *4GL* languages and presented low *productivity* values. On the other hand applications developed in *PC* was designated mainly for *communication* and *electricity, gas, water* and *energy* providers, were developed in *Access* and *SQL* and had high *productivity* values. Intermediate productivity values were observed in projects developed in *SQL*.
- Regarding the *size* of the projects measured in Function points it appears that it is negatively correlated with *the use of case tools*.
- It seems that productivity values remain equally distributed into intervals and unaffected by the time. Someone would expect that in most recent projects productivity would be higher but this is not justified by the analysis. It seems that IT personnel becomes more experienced as time passes but still they have to practice in a variety of new technologies.

The results of ABE pointed out that the best configuration to use is Canberra distance metric [2] with one analogy, considering as a point estimate the mean value of the estimated interval. Canberra distance metric is defined as following:

$$d_{new,i} = \left\{ \sum_{j=1}^{k} \frac{|Y_i - X_{ij}|}{|Y_i + X_{ij}|} \right\} \quad i = 1, 2, ..., n \tag{3.3}$$

The best attribute subset used for estimation is considered the *development platform* and the *development type*.

3.7 Identification of Associations Between Attributes

In this section we will identify several useful patterns that exist in the IS project data. Frequent and pertinent relationships among the project attributes are discovered and interpreted intuitively. For this purpose AR will be utilized as a method for descriptive modeling. With rules as those presented in Table 3.11 it is possible to identify often used techniques and methodologies, trends in the development of IS and their evolution over time and finally patterns that a practitioner should avoid.

Certain conclusions that we can reach from the rules are summarized in the following list:

- *Methodology* is used mainly in *Decision Support Systems* projects, developed inhouse, unlike *Knowledge Based* systems where no methodology is mostly used. The use of case tools is observed mainly in applications destined for *personnel business area type*.

Table 3.11. Association rules detected for attribute estimation

Support	Confidence	Rule Body	Rule Head
30.8	80.0	DSS+DevelopedInhouse	Methodology_Yes
43.6	100.0	4GL	DSS
25.6	90.9	MTS < 2.5	DSS
28.2	91.7	Communication	CaseTool_No
23.1	75.0	Communication	Methodology_No
25.6	83.3	Communication	KB
15.4	100.0	CaseTool_Yes+Methodology_Yes	DevelopedInhouse
10.3	100.0	BAT_Personnel	CaseTool_Yes
17.9	77.8	CaseTool_No+Methodology_No	KB
25.6	90.9	KB	CaseTool_No

- From 1993 to 2000 *communication* organizations demanded mainly knowledge based systems.
- The *4GL*, non AI languages used for the *New development* projects is mainly *Access* and *SQL*.
- The absence of *case tools* and methodology is mainly observed in the development of *knowledge based* systems.
- Projects developed by small *teams* < 2.5 and in *4GL* languages are usually decision support systems.

3.8 Conclusions and Future Work

In this chapter we examined four machine learning techniques in estimating the productivity for the development of IS. All of the methods were able to classify the majority of projects in the correct productivity interval and provide several assumptions regarding the development environment. CART are able to provide a simple, easily understood estimation framework, rules can provide a more informed estimation and BBN support thorough analysis of the dependencies and independencies among the projects attribute and productivity. Possible advantages of these methods in the estimation of IS systems are:

- The knowledge extracted from the application of the methods is relevant to the data collected. Therefore each organization can specify the data of interest, collect the appropriate information and estimate the target fields. Therefore even if in our study the data analyzed were of multi organizational nature a company could calibrate and adjust the models to its needs collecting the appropriate data.
- The output of the models is easily understood. The models support visual representation of the results, have a strong mathematical background and can be extracted with the use of several tools available free from the internet [9, 37].

It is evident that for more detailed analysis of these issues further research has to be done applying and comparing the previously mentioned methods on data involving IS development with pure AI techniques as well. Data pertaining to the development of IS could involve the mining algorithm used for the implementation of them, the tools, languages and platforms utilized specifically for the development of such projects, the requirements in storage and memory during the application of the systems. When a large cost data base for AI development projects is available, more thorough analysis will be possible and the pros and cons of each development tool, technique or algorithm will be easier to observe. In addition sensitivity analysis will show the robustness of the methods. Finally, collecting such data over long periods will permit trend analysis, providing a picture of AI development evolution over time.

References

1. Albrecht, A.: Measuring Application Development Productivity. IBM Application Development Symposium, Monterey California, October (1979) 83–92
2. Angelis, L., Stamelos, I.: A Simulation Tool for Efficient Analogy Based Cost Estimation. Empirical Software Engineering, 5(1) (2000) 35–68
3. Bibi, S., Stamelos, I., Angelis, L.: Bayesian Belief Networks as a Software Productivity Estimation Tool. Proc. 1st Balkan Conference in Informatics, Thessaloniki Greece (2003) 585–596
4. Bibi, S., Stamelos, I., Angelis, L.: Software Productivity Estimation Based on Association Rules. Proc. of 11th European Software Process Improvement Conference, Trondheim Norway (2004) 13A1
5. Bibi, S., Stamelos, I., Angelis, L.: Software Cost Prediction with Predefined Interval Estimates. Proc. of 1st Software Measurement European Forum, Rome Italy (2004) 237–246
6. Boehm, B.: Software Engineering Economics. Prentice-Hall, Englewood Cliffs NJ (1981)
7. Boehm, B., Abts, A., Brown, W., Chulani, S., Clark, B., Horowitz, E., Madachy, R., Reifer, D., Steece, B.: Software Cost Estimation with COCOMO II. Prentice Hall, Upper Saddle River NJ USA (2000)
8. Breiman, L., Friedman, J., Oshlen, R., Stone, C.: Classification and Regression Trees. Wadsworth International Group, Pacific Grove CA (1984)
9. Chapman, P., Clinton, J., Kerber, R., Khabaza, T., Reinartz, T., Shearer, C., Wirth. R.: CRISP-DM 1.0 steb-by-step data mining guide. http://www.crisp-dm.org/Process/index.htm CRISP-DM (2000)
10. Chen, S., Liao, C.: Agent-Based Computational Modeling of the Stock Price Volume Relation. Information Sciences, 170(1) (2005) 75–100
11. Cheng, J., Greiner, R., Kelly, J., Bell, DA, Liu, W.: Learning Bayesian Networks from Data: An Information-Theory Based Approach. The Artificial Intelligence Journal, 137 (2002) 43–90
12. Domingos, P.: Metacost: A General Method for Making Classifiers Cost-Sensitive. Proc. Knowledge Discovery in Databases (1999) 155–164
13. Erman, L., Lark, J., Hayes-Roth, F.: ABE: An Environment for Engineering Intelligent Systems. IEEE Transactions on Software Engineering, 14(12) (1988) 1758–1770

14. Hand, D., Mannila, H., Smyth, P., Principles of Data Mining. MIT Press, Cambridge, MA (2001)
15. International Software Benchmarking Group: www.isbsg.org
16. Jeffery, R., Wieczorek, I.: A Comparative Study of Cost Modeling Techniques Using Public Domain Multi-Organizational and Company-Specific Data. Proc. European Software Control and Metrics Conference, Munich Germany (2000) 239–248
17. Jeffery, R., Ruhe, M., Wieczorek, I.: Using Public Domain Metrics to Estimate Software Development Effort. Proc. 7th IEEE Int. Software Metrics Symposium, London UK (2001) 16–27
18. Jensen, F.: An Introduction to Bayesian Networks. Springer, Berlin Heidelberg New York (1996)
19. Jorgensen, M.: A Review of Studies on Expert Estimation of Software Development Effort. Journal of Systems and Software, 70(1–2) (2004) 37–60
20. Jorgensen, M.: Practical Guidelines for Expert-Judgment-Based Software Effort Estimation. IEEE Software, 22(3) (2005) 57–63
21. Jorgensen, M., Sjøberg, D.: An Effort Prediction Interval Approach Based on the Empirical Distribution of Previous Estimation Accuracy. Information and Software Technology, 45(3) (2003) 123–126
22. Kleinberg, J., Papadimitriou, C., Raghavan, P.: A Microeconomic View of Data Mining. Journal of Data Mining and Knowledge Discovery, 2(4) (1999) 311–324
23. MacDonell, S., Shepperd, M.: Combining Techniques to Optimize Effort Predictions in Software Project Management. Journal of Systems and Software, 66(2) (2003) 91–98
24. Mair, C., Shepperd, M.: An Investigation of Rule Induction Based Prediction Systems. Proc. Int. Conf. on Software Engineering, Los Angeles USA (1999)
25. Mair, C., Kadoda, G., Lefley, M., Phalp, K., Schofield, C., Shepperd, M., Webster, S.: An Investigation of Machine Learning Based Prediction Systems. Journal of Systems and Software, 53(1) (2000) 23–29
26. Marbán, O., Amescua Seco, A., Cuadrado, J., García, L.: Cost Drivers of a Parametric Cost Estimation Model for Data Mining Projects (DMCOMO). ADIS Workshop on Decision Support in Software Engineering, Madrid Spain (2002)
27. Maxwell, K., Applied Statistics for Software Managers, Prentice-Hall, Englewood Cliffs (2002)
28. Mirhaji, P., Allemang, D., Coyne, R., Casscells, W.: Improving the Public Health Information Network Through Semantic Modeling. IEEE Intelligent Systems, 22(3) (2007) 13–17
29. Passing, U., Shepperd, M.: An Experiment on Software Project Size and Effort Estimation. Proc. Int. Symp. on Empirical Software Engineering (2003) 120–127
30. Puppe, F.: XPS-99: Knowledge-Based Systems – Survey and Future Directions, Lecture Notes in Computer Science of the 5th Biannual German Conference on Knowledge-Based Systems, Würzburg Germany (1999) 193–200
31. Rissanen, J. Stochastic Complexity in Statistical Inquiry, World Scientific Publication, River Edge (1989)
32. Ruhe, M., Jeffery, R., Wieczorek, I.: Cost Estimation for Web Applications. Proc. 25th Int. Conf. on Software Engineering (2003) 285–294
33. Sentas, P., Angelis, L., Stamelos, I.: Software Productivity and Effort Prediction with Ordinal Regression. Journal of Information and Software Technology, 47(1) (2005) 17–29

34. Stamelos, I., Angelis, L., Dimou, P. Sakellaris, E.: On the Use of Bayesian Belief Networks for the Prediction of Software Productivity. Information and Software Technology, 45(1) (2003) 51–60
35. Turban, E., Aronson, J., Liang, T., Sharda, R.: Decision Support and Business Intelligence Systems. Prentice Hall, Upper Saddle River (2006)
36. Virvou .M: A Cognitive Theory in an Authoring Tool for Intelligent Tutoring Systems. IEEE Int. Conf. on Systems Man and Cybernetics 2002, Vol. 2 Hammamet Tunisia (2002) 410–415
37. WEKA: http://www.cs.waikato.ac.nz/ml/weka
38. Welzer, T., Rozman, I., Druzovec, M., Brumen, B.: Reusability and Requirements Engineering in Intelligent Systems. IEEE Int. Conf. on Systems Man and Cybernetics, Vol. 5 (1999) 796–799
39. Witten, I., Frank, E.: Data Mining: Practical Machine Learning Tools and Techniques. Morgan Kaufmann, San Francisco (2005)

4

In Search of Narrative Interactive Learning Environments

Paul Brna

The SCRE Centre, University of Glasgow, St Andrew's Building, 11 Eldon Street, Glasgow G3 6NH, Scotland, paul.brna@scre.ac.uk; http://www.scre.ac.uk/personal/pb/

Summary. The promise of Narrative Interactive Learning Environments is that the various notions of narrative can be harnessed to support learning in a manner that adds significantly to the effectiveness of the learning environment. The task that this chapter addresses is the identification of the paths along which researchers need to travel if we are to make the most of the insights that are currently available to us.

4.1 The Promise of Narrative

In the last fifteen years, there has been a growth of interest in interactive learning environments[1] to which the designers have added the word 'narrative'. For the purposes of this chapter we define a 'Narrative Interactive Learning Environment' (NILE) to be an environment which has been designed with an explicit awareness of the advantages for learning to be cast as a process of setting up a challenge, seeking to overcome some obstacles and achieving a (partial) resolution.

Since many, if not all, interactive learning environments (ILEs) can be cast in terms of challenge, obstacle and resolution, we seek to explore the advantages (and disadvantages) of viewing the enterprise of ILE design as essentially NILE design. Our approach here is to present some of the key concerns of those researching in the area of NILEs with the help of a simple framework that focuses on learning. In doing so, it is hoped that the reader will begin to see NILEs in a new light.

Part of this hope is that the reader will also come to understand how the term 'narrative' is to be understood in the context of NILE design,

[1] The term *interactive* is used to bring the notions of narrative and interaction into a productive tension. While many of the systems that are of interest are also *intelligent*, the interest here is primarily in the ways in which learners progress their understanding and develop their skills. The implications for intelligent learning environments should 'fall out' of the analysis.

P. Brna: *In Search of Narrative Interactive Learning Environments*, Studies in Computational Intelligence (SCI) **104**, 47–74 (2008)
www.springerlink.com

development and deployment. Further, there is the question to answer as to whether ILE designers ought to be explicitly aware of, and trained in, the uses of narrative in design.

One reason why the notion of a 'Narrative Interactive Learning Environment' is attractive to some is the connotation with the potential for stories to engage the reader. There seems to be an implicit promise that NILEs will have, for example, the intensity of seeing an exciting film, or reading an absorbing book. Also, because of the association with the purpose of ILEs as promoting effective learning, there is also the suggestion that the learning experience will be successful.

From the perspective of those engaged in developing increasingly interesting and engaging narrative experiences, the aims are clear.

> "The player is free to move around the world, manipulate objects, and, most importantly, interact with the other characters. But all this activity is not meandering, repetitious, or disjoint, but rather builds in compelling ways towards a climax and resolution. Thus, in an interactive drama, the player should experience a sense of both interactive freedom and a compelling, dramatic structure" [1, p. 7]

From the perspective of those looking to improve learning, designers want their applications to engage the learner, and have sought to utilise notions associated with narrative [2, 3]. The emphasis can sometimes draw on the related issue of how dramatic performance can be used for pedagogic purposes [4,5].

So the enterprise in this chapter is to take two rather diffuse notions – that of a NILE and that of narrative experience – and explore the ways in which the two notions illuminate each other. To help us do this, we will provide a framework for analysing NILEs. We will outline some of the ways designers have realised their understanding of narrative, and then briefly consider the evidence in favour of the narrative approach. Finally we discuss possible future developments.

4.2 A Framework

With narrative having such a pervasive role for all aspects of life, a comprehensive framework is inevitably quite complex. What simplifies matters a little is that we are primarily interested in the role of narrative in learning. We classify the uses of narrative into five primary contexts that relate to the main purpose of learning.

Preconditions for learning: Arranging for the learner to be prepared for some learning experience in the fullest possible manner. This preparation can include cognitive, affective and conative factors, as well as situational and social ones.

Postconditions for learning: Ensuring that appropriate activities take place that encourage the learner to consolidate their learning, maintain or increase their motivation, reflect on their experience and so on.

Learning content: The actual interaction with material to be understood, mastered and applied.

Design: Prior to any learning there is the planning that is needed to set in motion the events that lead to the experience. Design involves the development of the underlying software, the kinds of interactions with the learner that are allowed, and the narrative within which the learner is to be placed.

Evaluation: Both during and after the learning experience there is a need to examine the learning experience with a view to making sure that the learner benefited. This process might well lead to suggestions for improving the experience.

While the above five contexts relate to the deployment of NILEs, the learner has to be considered in terms of what they seek from learning in general, and from NILEs in particular.

People can be (self) motivated towards achieving many goals. For the purposes of this chapter we consider five general aims that people might have when working with a NILE. These aims are neither mutually exclusive nor independent. Nor are they necessarily the 'designed-in' aims of any specific NILE.

Competencies: becoming more knowledgeable/skilled – the 'classical' view of the purpose of learning for the learner who may want to be able to speak a second language, repair a car etc.

Creativity: becoming more creative – one of the aims associated with broadening interests. Learners go to art classes, attend design courses etc.

Self Identity: improving one's sense of personal identity – the learner's aim is to become more confident, believe in themselves, cope better with personal crises.

Social Identity: improving one's social identity – learners want to receive the approval of others, to please their friends, to be the centre of a group and so on.

Relationships: improving relationships – learners want good friends, good role models and good relationships with their parents, teachers etc.

The issue of whether the goals/aims are 'designed in' is important for this chapter. We can analyse learners in terms of the complex set of aims that they hold both habitually and from time to time. This analysis would naturally illuminate the flow of actions and interactions that take place when the learner uses a NILE. Indeed, we could do the same for any learner and any interactive learning environment. However, we would not necessarily be the wiser for what makes the difference between a NILE and any other interactive learning environment. The method of examining the learner's aims in relation to the use of an ILE/NILE is useful, but what we want to know is what difference narrative makes to an ILE.

While there may be some reason to believe that NILEs should be better than an ILE, there is no compelling reason to believe that this happens in practice. Once upon a time it was believed that intelligent learning environments would be more effective than (unintelligent) learning environments. Thirty years later, we are still seeking understanding of the benefits of adaptive systems [6]. The situation in relation to NILEs is similar – we need more good examples, and we still need to understand much about what benefits NILEs bring. The issue about the benefits (and costs) of taking the NILE path rather than the ILE path is an open one. However, even if the costs turn out to be the determining factor, there is one kind of benefit that is clear from what follows – NILEs can help designers refocus their attentions upon the whole person.

4.2.1 Preconditions for Learning

Preparing the learner is an important idea. Gagne, in his work on the conditions of learning [7], argued that a new skill requires that the learner has already mastered the skills upon which the new skill depends. Similarly with concepts, the learner should know about mass and velocity if they are to learn about momentum. Even this apparently self evident position is by no means as certain as it might appear. Raven showed that children were able to cope with the concept in an informal and 'global' manner [8].

Recalling Ausubel's notion of advance organisers [9], we note that Ausubel was not just concerned with preparing the learner to learn by providing the vocabulary to be used, or even making sure that certain preconditions were met [10].

> "Advance organisers bridge the gap between what the student already knows and the new material at a higher level of abstraction, generality and inclusiveness." [9]

Advance organisers are more concerned with intellectual preparation. Other approaches have been suggested, and used, for motivational and attitudinal aspects of learning – most famously, Keller and his colleagues [11,12] who introduced the ARCS model involving attention (A), relevance (R), confidence (C) and satisfaction (S). The ARCS model is used by many to support the design of ILEs that are intended to engage learners.

Narrative is also used to set up the situation in which learning takes place. This is, in a deep sense, a precondition for the use of narrative – and most situations in which it is intended that learners learn something. From a situational perspective, setting up the social and physical context in which action will happen requires some engagement with the concerns that are central to narrative design – whether or not this is acknowledged explicitly.

Perhaps the most natural usage is in the exposition of the 'back story' for each of the 'actants' in the situation [13]. Generating such a narrative is by no means straightforward but, if done well, can provide not only the context but also help to motivate learners. It is also possible to convey some of the notions that are closely related to Ausubel's advance organisers.

4.2.2 The Postconditions of Learning

While Ausubel is well known for his notion of advance organisers, the idea of post organisers is less familiar, yet there is evidence as to their effectiveness [14]. Hall, Hall and Saling examined the effectiveness of knowledge maps – a form of concept map – when used as a post organiser. Their empirical work suggested some advantages to this approach. In particular, participants who engaged in a postorganisational activity with the help of a knowledge map structure (i.e. with text removed) recalled significantly more of the information than those that used the postorganisation time to continue to study the material to be remembered. The effect seems to be stronger with superordinate propositions. Further work suggested that the effect was also stronger for those who did postorganisational work with the full map (i.e. with more information present).

Much of the recent work that covers related ideas have tended to draw on Schön's notion of reflection-on-action [15]. In the last few years there seems to have been a noticeable increase in interest in systems that encourage the learner to reflect on their learning or on the way that they learn.

4.2.3 Learning the Content

Such a use of narrative might embed the coming task/situation into a suitable story – real or imaginary. For example, Ariadne's thread is a classic use of a fantasy story used to support a number of challenges to the learner with the aim that they learn about logic programming [16]. Waraich's Binary Arithmetic Tutor (BAT) system is another example of a narrative in which tasks connected with binary arithmetic are embedded [17]. His MOSS (Map reading and Ordinance Survey Skills) system, developed as a pilot for his approach to narrative centred informant design, was aimed at teaching map reading skills to children [17]. The MOSS system weaves the narrative in together with the tasks.

The basic issue is how to make the narrative aspects of an ILE work for the learning aims and goals, and how to ensure that the learning goals do not undercut the narrative aspects in such a way that the whole endeavour is compromised.

4.2.4 Design

Traditional methods can be used for the design of NILEs. Learner centred design [18] has been increasingly favoured; there have been some notable

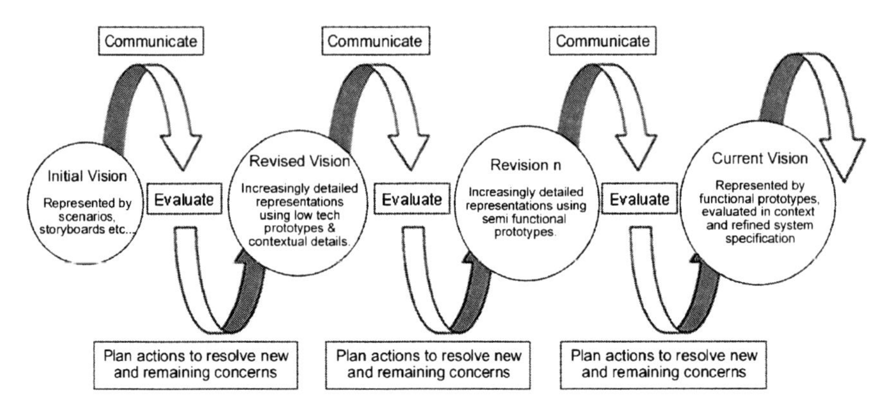

Fig. 4.1. Luckin's HCD process from [22]

developments in the approaches taken. These include Scaife, Rogers, Aldrich and Davies in their development of informant design [19] as well as Chin, Rosson and Carroll's scenario-based design [20] and illustrated by Cooper and Brna in their work on the development of T'rrific Tales [21].

A recent special issue of the International Journal of Artificial Intelligence (Vol. 16, number 4) provides a good overview of the ways in which learner centred design is conceived as a general means of developing ILEs. In particular, Luckin, Underwood, du Boulay, Holmberg, Kerawalla, O'Connor, Smith and Tunley provide a detailed account of their learner centred methodology for software development [22]. They term their approach an example of *Human Centred Design.*

Their approach seeks to identify the stakeholders and then involves them working with the stakeholders using, for example, storyboards and interviews. The process is cyclical and represented in Fig. 4.1. It is claimed that the experience of working through this cycle leads to an increasingly rich understanding of the needs of the learners.

4.2.5 Evaluation

If we want to know whether a particular NILE is delivering the goods then we need to evaluate the current state of the design in relation to the kinds of people who have an interest in the outcomes.

Since NILEs work on so many levels – cognitive, affective and conative, as well as on self-identity and personal relationships – the methods needed for any evaluation are very varied. We are not 'simply' looking for learning gains on standardised tests, we are also looking for more elusive gains. Self identity, for example, is something that can be examined throughout one's life, and there are no simple metrics that can identify changes in some absolutely 'right' direction. However, almost everything is amenable to evaluation at some level of detail, with some degree of plausibility.

In the case of NILEs we will find both standard methods from experimental psychology, and ones that are qualitative . While there is no obvious requirement to evaluate the effectiveness of a NILE using 'narrative' methods, there is a place to use methods that can loosely be described under the heading of 'narrative inquiry' [23] which seeks to take the stories of participants very seriously.

4.3 The Achievements

We select some clear exemplars that demonstrate distinctive qualities for the first four primary contexts – but the fifth primary context, that of evaluation, does not have any strong exemplar. However, some promising approaches are outlined.

4.3.1 Ghostwriter – Preparing the Learner to Write

As an example of the use of narrative to prepare the learner to take part in some specific learning experience, we examine Robertson's thesis work on Ghostwriter [24]. Ghostwriter was based on the idea that children who find it difficult to write stories could be provided with a stimulating experience which would then provide them with the germs of ideas that they could turn into stories. Ghostwriter is a 3D environment designed to encourage role playing by two participants who help each other to complete the task given to them [24].

The environment is one in which the two players are forced to make ethical decisions and form judgements about other characters in the game [25]. There is a human facilitator who makes choices to keep the role playing game 'on track'. The game was designed to avoid the point and shoot style of many virtual environments in order to encourage imaginative characterisation and plotting.

After the children had discussed their experiences, they were encouraged to write a new story. The empirical evidence obtained from this work is impressive: learners were motivated, developed good working relationships with each other, identified with the characters in the game [24] and, importantly, their new stories featured a greater use of the relationships between characters [26].

The 'Ghostwriter' scenario is a clear use of a NILE to give learners an experience to prepare them for an educational task. The designer's aim with Ghostwriter may have been that it should be a preparatory experience that supports the development of a learner's story writing skills, but the experience-in-itself seems to have been a success with both children enjoying its use and teachers perceiving benefits of educational value such as improving self-esteem and their classroom behaviour [24] .

One question that arises – are there any useful examples of preparation for tasks other than story telling ones? Certainly there is evidence that this can be done. Boström, in her thesis work used a narrative inquiry method to examine student narratives [27]. For example:

> "Gustaf, 30 years of age, was interested in motorcycles and had experienced corrosion problems when screws made from stainless steel were used in engine blocks made from aluminum. He told me about the world of making 'choppers', motorcycles stripped from most of the gear. His fascination for motorcycles and chemistry's capacity to explain material problems was the explanation for why he changed from being a student that hardly managed into a student making excellent results, progressing into becoming an engineer, studying material technology."

While it is difficult to engineer a life changing experience, Gustaf's own experience indicates the potential that a NILE could have in exposing a student to a story which provides phenomena that can be the basis for pursuing a scientific explanation, and, eventually, come to comprehend the underlying science. Not only does a NILE have the capacity to ground scientific inquiry in real life situations, it can also motivate the learner to learn the underlying domain (Gustaf also was highly motivated by his experience).

There are many ILEs that perhaps can be reconceived as NILEs following a similar path – i.e. providing the phenomena in a realistic situation that needs an explanation, engaging the student in a situation which motivates them to seek an answer. Even the earliest ILEs often presented some simplified view of the world in a manner that could stimulate learning. For example, ROCKET [28] provided an experience of managing a rocket in space with little or no net gravitational force upon it, and O'Shea created a range of microworlds that allowed the exploration of science phenomena [29].

Such environments may have provided a back story, and motivated the learner through presenting counter intuitive events – but such environments do not have all the features they might have. While the learner is part of such environments, controlling the action – and often being surprised by the consequences, the context is 'stripped' of much of what might convince the learner that the tension between what was expected and what actually happened was real.

Much more recent efforts have been made to embed learning in some realistic context. For example, Barab, Sadler, Heiselt, Hickey and Zuiker have sought to embed learning within a multiplayer gaming environment called Quest Atlantis [30]. The environment appears to motivate the students but there are difficulties connected with managing the learning experience.

Environments such as ROCKET or Quest Atlantis are not 'stand alone' ILEs. Whether the environments act to prepare the ground for learning, or as part of an approach to remediation, they are embedded in the learner's experience by the way they are used by teachers – even if there are significant levels of negotiation that take place between learners and teachers. They are located within a social context in which there is a polylogue focused on learning. It would seem evident that NILEs are not intended to be stand alone but provide support for learning on the assumption that they are deployed within a context focused on learning.

4.3.2 FearNot! – Improving Social Relationships

While there are many ILEs that are designed for learning science and mathematics, modern NILEs featuring role playing immersive environments are often targetted at procedural training, topics in the humanities or connected with social and psychological aspects. This latter case is of great value with a growing awareness in some countries of the urgent need to socialise young people into good relationships with each other, older people and various social institutions (not least, the schools themselves).

While some environments such as FearNot! (see below) focuses on personal experiences of bullying and the development of ways in which bullying might be managed, others have looked at the ways in which people can seek to grow/restore their sense of personal worth and relationships with others [31]. Their emphasis is on telling stories which evoke connections with the learner, and give insights into their own personal circumstances. This is a powerful use of narrative, but how does it fit with the emerging conception of a NILE?

Of the making of stories there is no end. Each learning experience may generate a story from which something is taken, memorised and learned. We do not need a NILE for this to happen – but a NILE might well facilitate this by encouraging the learner to 'tell' their own stories whether within the NILE or elsewhere.

FearNot! is one of the most significant NILEs of recent years. The VICTEC (Virtual ICT with Empathic Characters) project which developed the FearNot! system is an approach to help children understand the nature of bullying and, perhaps, learn how to deal with bullies. The system was targeted at 8–12 years old in the UK, Portugal and Germany. When using FearNot! the child views a sequence of incidents that features direct bullying. The characters in the scenes are synthetic, and the overarching context is a school.

The characters needed to be believable because it was intended that children should be engaged with the situations, and care about the outcomes for the characters involved. The intention was that the children using the system should both empathise with the characters and help them through proffering advice between incidents. This advice affects the synthetic character's emotional state which in turn affect the actions the character selects in the next episode. Hence in two ways, the child is intended to be a 'safe' emotional distance from the incidents – through an empathic relationship (rather than one in which the child identifies directly with the characters – see [32]), and by trying out ways of dealing with bullying through offering advice [33,34].

4.3.3 SAGA – Planning and Managing the Experience

Some of the most influential researchers who have sought to produce an engaging interactive narrative experience have not been particularly interested in learners. Their focus has been more on the intriguing problem of how to reconcile the notion of narrative as found in, say, literary fiction and the sense of indeterminacy that comes with unconstrained interaction.

The argument about planned versus emergent narrative is important for those interested in interactive learning environments – but perhaps only up to a point. Within the context of designing an ILE, there is no doubt that there needs to be some knowledge that the learner is to gain, some experience that is valued ... usually, a well articulated purpose even if it is difficult to be precise about what exactly will be learned. There also needs to be a clear awareness of education as essentially a process that takes place through substantive interactions between learner and teacher or learner with peers. So, any approach that combines planning aspects of the learner's experience and giving the learner (and the ILE) some scope for constructive interactions has a potential worth considering.

Key early work in the area of managing events within a narrative experience was undertaken within the Oz project [1]. The most relevant work from that project is arguably that of Weyhrauch who developed a drama manager [35], an idea described by Laurel [36]. Weyhrauch's achievements were mainly at the technical level of guiding the developing interactive narrative. Arguably, his main achievement lay in finding evaluation functions with which to score the possible future moves to see which sequence was most promising.

Another of the Oz group, Mateas, is responsible for the Façade interactive drama [1]. His thesis added a new way of managing interactive dramas by focusing on the theory of dramatic beats.

> "But how should beat behaviors be structured? By using the term 'beat', one begins to tap into the rich meanings that surround the word 'beat' in dramatic theory. In dramatic beats, story values change over the course of the beat. The story values change in response to the within-beat actions and reactions of characters. This suggests that specific sub-beat actions must be communicated to the audience in order to accomplish the value change. This, combined with the ways of talking about goals and behaviors in ABL, suggests the idea of beat goals – sub-beat units that must be accomplished in order for the beat to happen. Of course beats in an interactive drama must respond to player interaction – somehow player interaction must be incorporated into the default beat goal structure." [1, p. 229]

Mateas explains that beats are blended into their surroundings using a technique that Sengers emphasised in her thesis [37] – transitional behaviours. Sengers pointed out, quite rightly, that while robots might move from action A to action B with little or no indication that a transition is happening, people interacting with others do not behave like this under normal conditions.

What they do is signal for the benefit of others what is about to happen. This transitional behaviour is important for believable agents in some learning environment – and perhaps as important for teachers in real classrooms. Mateas gives authors of his beats tools that help them build transitions.

> "In our idiom for Façade, beats generally have a transition-in beat goal responsible for establishing the beat context and possibly relating the beat to action that happened prior to the beat, a transition-out beat goal communicating the dramatic action (change in values) that occurred in the beat as a function of player interaction within the beat, and a small number of beat goals between the transition-in and transition-out that reveal information and set up the little interaction game within the beat." [38]

Mateas' dramatic beats are like mini-narratives. So we might conclude that one aspect of a NILE is that transitions are an important aspect. The broader lesson is that the author of a NILE needs tools that allow them to add narrative structure easily. The more important issue here is to assess whether such an approach really does aid in the management of the learning experience.

Generally, ILEs lack any explicit management of the total shape of the interaction. Many good systems have some structure hard wired into them – the MENO project examined the effects on the learner of providing various kinds of structure in the design of the environment [2]. Additionally, many adaptive systems have been designed to select content 'on the fly' – for example, the DCG system [39]. Until relatively recently, few sought to do more than estimate the next piece of content to present based on progress through the content. With the rise of interest in managing motivation, we now see systems that seek to take affective and motivational factors into account when choosing content.

Del Soldato and du Boulay's MORE system was an early and ground breaking system that used a simple method for deciding whether or not to select more challenging material [40]. Murray's DT Tutor is capable of making decisions about which of several possible tutorial actions is most likely to be of benefit from a decision-theoretic perspective [41]. To do this, it takes into account the student's knowledge as well as affective state and task progress.

However, there are still few systems that set out to manage the total experience in even a limited manner. Work by Szillas, Machado and Young does address planning for a narrative experience while still allowing for individual freedom. However, much of this work is only managing the narrative experience and does not really examine whether a NILE is useful for teaching Physics, Assessment, French etc. Szilas, like Mateas, is primarily interested in interactive drama [42, 43] – i.e. the questions that concern him are more connected with whether the experience is suitably entertaining.

Figure 4.2 shows the way that Szilas sees IDtension from an architectural point of view. The notion of using such an architecture for making an

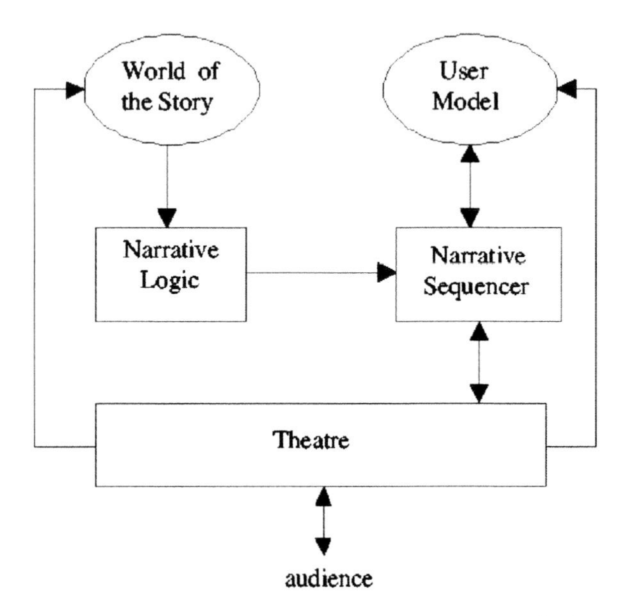

Fig. 4.2. Szilas' IDTension architecture from [43]

interactive drama out of, for example, a chemistry experiment is not unreasonable but the main issue may be just how convincing a game can be generated from the domain. If a good chemistry game is possible then Szilas' (or Mateas') approach should work well. So there is a clear gap here between the world of narrative/drama and the world of chemistry. The gap suggests that we need methods for designing appropriate content-based educational games. The current 'serious games' movement is another attempt to harness the interest in games for 'serious' purposes such as found in education and training [44, 45]. The problem is a significant one, however. For example, Mateas and Stern point out that authoring of their Façade system is very demanding [46], and this is suggestive of the difficulties faced by those building narrative oriented environments. We also have the the problems of building good content oriented authoring tools [47], and building authoring tools for games.

The problems taken together with the achievements suggest that progress in providing tools to accelerate the process of building NILEs has been quite slow and piecemeal.

Young and his colleagues have worked on an extensive, and impressive, range of problems connected with producing more engaging narratives under some degree of automated control. His Mimesis architecture builds a story plan derived from all the possible actions that can be generated by the agents and the user in the scenario [48, 49]. The popular Hierarchical Task Network (HTN) approach is used for planning.

While Young's interests are not primarily educational, Riedl and Young have demonstrated how their thinking might be applied at school level [50].

Riedl's Fabulist architecture is applied to a scenario in which a learner controls a character in a historical simulation. This particular approach to history education has been popular from time to time. Though it represents only one approach to learning history, it is good to see such an example of a NILE.

As mentioned previously, Marsella, Johnson and colleagues at the Center for Advanced Research in Technology for Education (CARTE) at the University of Southern California have made a significant contribution to the notion of 'interactive pedagogical drama'. Their work has a specific aim to train people to deal with situations that are, on the whole, inherently stressful. Carmen's Bright IDEAS is intended to support the development of problem solving skills for mothers of paediatric cancer patients [5]. Learners influence Carmen by interacting with the system which has the effect of distancing the learners from their own situation while still focusing on learning useful skills. This approach is similar to the one adopted for the VICTEC project. The evaluation outcomes in relation to the pedagogical goals seem to be encouraging.

Machado's Support And Guidance Architecture (SAGA) was a significant attempt to develop a more learner-centred support architecture for collaborative learning [51–54].

The aim of the work was to produce a kind of plug in component that could be included in a variety of systems designed for story creation. It has been tested with Teatrix, a 3D collaborative story creation environment designed for eight year olds [55]. Perhaps the most significant educational innovation was the inclusion of a method for encouraging reflection through 'hot seating', derived from the approach developed by Dorothy Heathcote to the use of drama in education [56].

The abstractions used by Machado to build a form of drama management are derived from Propp's work on narrative analysis [57]. Machado used the 31 functions described by Propp and the seven roles of villain, hero, donor, helper, princess (and father), dispatcher, hero and false hero. She uses the structuralist view of narrative [13,58] to set up the main actions but does not ignore the need to develop believable characters with which the learner may identify.

In order to integrate SAGA with other applications, Machado developed a communication protocol. Her architecture is shown in Fig. 4.3.

In addition to the narrative guidance engine, SAGA's components includes a facilitator, a script writer, a director agent and the reflection engine.

The narrative guidance engine generates the space of all plot points which are derived from the character's roles, the initial story situation and achieving the specified goal [59]. As with Weyhrauch, there is an evaluation function that seeks to predict which path is most satisfying with an emphasis on improving the learner's concept of narrative and giving them a good experience of collaboration with the other learners using the system.

The reflection engine is the component that generates a 'reflection moment' consisting of a request for the learner to stop their work in the learning

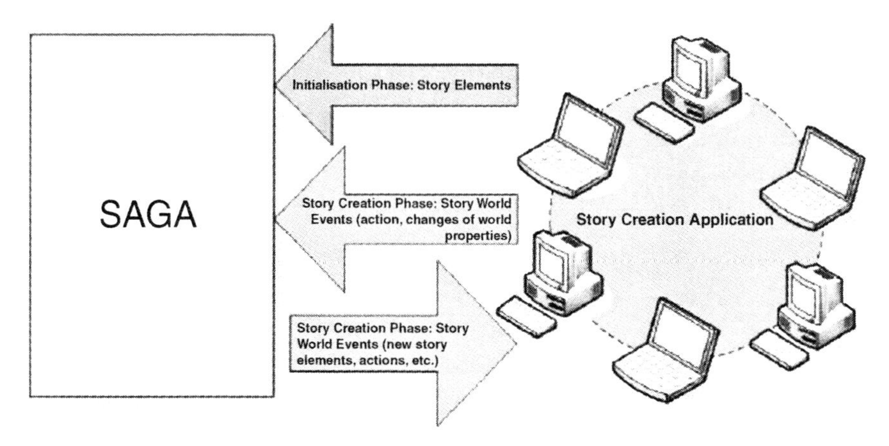

Fig. 4.3. Machado's SAGA architecture from [51, p. 148]

environment and review their character's behaviour and the story's development. Heathcote makes it clear that such a move in the classroom is more than just a means of generating reflection. She sees this as a *failure saver* as well as a *slower down into experience* [56].

In this section we have looked at what has been achieved with respect to planning and managing the experience. We have discussed a variety of drama managers. What we have found is very little that could be regarded as a programmatic attempt to blend the work on narrative with that of ILEs. However, this issue has been addressed by Riedl, Lane, Hill and Swartout at the Institute for Creative Technologies, University of Southern California [60]. While the goal is excellent, and the analysis of the tensions between the overall narrative experience and standard intelligent tutoring is valid, this is neither sufficiently well developed nor the only way that one might conceive of the ways in which we might hope narrative and learning experiences might be managed.

One area of planning narrative experiences which seems to have been highly productive is that exemplified by Machado's reflection tool which emphasises the role of the learner to engage with the narrative. While this approach takes one out of the narrative in order to reflect on it, it is this that makes Machado's work attractive. There are two major pathways – 'narrative as motivation' and 'narrative as the ways in which we approach difficult ideas and experiences'. Machado uses both approaches, but it must be clear that, in the situation in which any NILE is used, the capability to move in and out of the engaging experiences and take stock of the situation in terms of what has been learned and how the experience can be built upon is at the heart of the educational uses of NILEs.

4.3.4 NCID – Bringing Narrative into the Design Process

One of the papers by Hall, Woods and Aylett, describes the development of FearNot! [61]. FearNot! is a system with strong narrative elements in its method of helping learners come to a better understanding of how to manage situations that involve bullying. The methods used for the design and development included working with a wide range of stakeholders. The approach is a form of informant design [19]. Informant design seeks to draw on the expertise of various stakeholders. When working with children in particular, it can be very difficult to extract the key information from the contributions being made. Informant design seeks to recognise the different kinds of contributions made by different contributors.

While there is no explicit reference to a methodology specialised to the development of NILEs, several of the components of their method are relevant to our interest in the design of NILEs. Hall et al. seek to provide a system which exploits many of the aspects of a designed narrative experience including storyboard generation, the creation of the events, characters and context in FearNot! and the story line.

Good and Robertson have developed a framework for participatory, learner-centred design involving children called CARSS [62]. CARSS stands for Context, Activities, Roles, Stakeholders and Skills; the approach seeks to make it much clearer how individuals work within the constraints to develop software. Again, the method is general – though it emphasises working with children, and, again, the method does not explicitly draw on any notions about narrative. However, as with Hall et al., Good and Robertson do address issues which are strongly related to the design of NILEs. In particular, their focus on roles played by members of the design team is of interest.

Waraich also takes an informant design approach in his work [17]. Although there are similarities between his narrative centred informant design (NCID) process [63] and those of Hall et al. (2004) and Good and Robertson (2004), there is a difference in that Waraich explicitly introduces narrative concepts into the process of structuring the contributions from the informants (Fig. 4.4).

Such an approach focuses on helping informants work on the problem of generating software which is engaging in terms of theme, setting, characterisation and plot structure. Providing informants with sufficient background in the understanding of narrative is challenging. Not only do different learners have different needs in order to participate constructively, but the need is conditioned to some extent by the curricular system in which learners grew up: the extent to which students learn 'narrative literacy' is a factor, and brings us to questions outside the scope of this chapter as to the state of knowledge about narrative within the various school systems around the world. There has been a growing awareness that literacy is more than the world of literature but there's a long way to go yet.

In terms of designing for NILEs, there is a further area worth mentioning – that of Empathic Design [64]. This is connected with the designer's

Design Phase	Methods	Contributors	Outcomes/Techniques
(1) Define domain and problems & identify basic narrative elements	Interviews Analysis of existing materials & software Talk about nature of narrative	Lecturers Domain expert Students Designer Narrative facilitator	User requirements document Learning objectives/goals Preliminary documentation produced Group discussions with focus group of students from years 2 & 3
(2) Translation of specification & plot/character definition	Storyboarding & sketching, scenario creation	HCI expert Software designer Narrative facilitator Students	Storyboards; index card based narrative description; Paper based narrative scenarios; Student's stories/group work Story analysis
(3) Design Low-tech materials and test	Storyboarding Scenario creation Design through scenarios plot definition; character matrices	Lecturers Domain expert Software designer Students Narrative facilitator	Storyboards; index card based narrative description; Paper based narrative scenario (cut outs)
(4) Design and test high-tech materials	Prototype (high-tech) Multimedia Pre-and post-tests Cognitive analysis	Psychologist Software designer Students HCI analyst Narrative facilitator	Prototype system (BAT) functional; Pre-& post-tests Questionnaires Observation Interviews

Fig. 4.4. Waraich's NCID process applied to construct the BAT system from [63]

duty of care to the learner. In the artificial intelligence in education community, John Self argued that caring for learners involved responding to their needs [65]. Caring for learners goes beyond effective and efficient learning of the specific content being considered, and looks to the wider picture, both in terms of the content and in terms of personal development. Designers of educational software need to factor empathy into the design process adequately, so that they can avoid concentrating too much on design issues connected with management and the curriculum.

4.3.5 Evaluating the Experience

As pointed out above, there is a need to be very flexible about the manner in which NILEs benefit learners. The approach needs to be suited to the kind of outcomes in which we are interested. For example, suppose we are interested in how engaged students are when using a NILE. Webb and Mallon used focus groups to examine engagement when students used narrative adventure and role-play games [66]. The method yielded some very useful guidelines which demonstrated some ways in which narrative devices could work within a game scenario.

Another interesting approach taken was by Marsh who, in his thesis work, turned the notion in VR of 'being there' on its head, and examined the notion of 'staying there' [67, 68]. For part of his work, he used Wright and Monk's method of cooperative evaluation to evaluate a desktop VR system [69]. He also went on to develop an evaluation method for three categories of experience – the vicarious, the visceral and the voyeuristic. The voyeuristic experience is associated with sensations, aesthetics, atmosphere and a sense of place, the visceral experience with thrills, attractions and sensations, and the vicarious as connected with empathy and emotional information [70].

The engaging experience that a NILE is supposed to provide needs such evaluations: the division proposed by Marsh is one way of categorising the kinds of experience that need to be evaluated. However, the approach needs to be fleshed out further – see [67] for how this approach was applied by Marsh.

Knickmeyer and Mateas have sought to evaluate the Façade system [71]. They were particularly interested in responses to interaction breakdown. Their method used a form of retroactive protocol analysis combined with a post experience interview; their analysis was based on a coding scheme developed from Malone's identification of the importance of challenge, fantasy and curiosity [72].

Laaksolahti and his colleagues have developed a 'sensual evaluation instrument' to measure some affective aspects of game experience [73]. The instrument is a set of objects that game players are asked to move around on a table after playing a game. Currently, the user moves the objects that are associated with experiences in the game closer to themselves (although this is work in progress). Another method they have been developing involves the familiar Repertory Grid Technique.

There is significant effort in the HCI community going into producing a range of methods for evaluating experience. The same is not so obviously true for those designing NILEs. Perhaps there are, as yet, too few such systems. Evaluating a NILE involves bringing to the evaluation the perspective of the pedagogue which tends to be missing from the evaluation of scenarios such as Façade. For such environments, the member of the audience is the final arbiter. If someone likes our product then they like it – (almost) end of story.[2] With pedagogy, the goals can too easily be 'imposed' on the learner, and the potential for liberating the learner using NILEs diminished. It will be no surprise to anyone that NILEs are like any other promising approach: their use depends to a great extent on the manner in which the system of education constrains the learner.

Evaluators of NILEs, such as there are, seem to find it hard to bridge the gap between evaluating experience and evaluating learning gains in relation to some content – or, perhaps, the problem has been that there is a lack of a plausible theory about the ways in which pleasure and learning are bound together.

---

[2] Not quite... there are always ethical issues in even the most apparently harmless of activities, and if something can be monitored, someone will want to do this.

The pedagogue may well favour learning gains, and the person developing edutainment systems the experience. There is a sense of balance needed here that evaluators have to attain. To fall into believing that instant gratification is primary is arguably a disaster for educationalists – but so is believing that content is primary to the exclusion of pleasurable experiences. Indeed, learning is often hard work, and much of it could scarcely be termed as pleasurable yet, at the end, with some learning goal attained, the learner may well look back and believe that there is great pleasure involved in learning. The evaluator needs to find some way of balancing the different forces at work (and these include more than pleasure and content). This is an activity for a mix of sociology, cognitive science, anthropology and education (at the very least) but an activity which also requires we take a view of learning (and education) that encompasses the widest possible range of needs that individuals have.

4.4 Where's the Pedagogy?

We need to mention the role of pedagogy[3] in the process of developing NILEs since NILEs are intended for learners, and the term *pedagogy* is associated with the teacher as the mediator of learning.

In many of the systems discussed, the underlying pedagogy is obscure – sometimes, because this is not seen as important by the authors/designers of the NILEs, sometimes because the pedagogy is taken for granted. On the other hand, some of the systems have a clear pedagogy in mind even if there are other ways of using the specific NILEs.

The VICTEC project had a clear pedagogy in terms of the way in which FearNot! was to be used in an experimental manner [34] but there is a need to consider the manner in which such a system is integrated into the normal curriculum. The experience of FearNot! needs to be 'wrapped around' with other constructive experiences and well chosen materials as well as changes to the wider school culture. The evaluation described in [34] makes it clear that there was no real follow up to examine the effects on learner's lives. It may be that the promised evaluation of the follow-up system, part of the ECIRCUS project, will provide some answers.

Another NILE, Carmen's Bright IDEAS, exhibits a potential for learning with an associated pedagogy. The term 'IDEAS' is an acronym for Identify a solvable problem, Develop possible solutions, Evaluate options, Act on plan and See if it worked [5]. The system was developed as part of the Bright IDEAS programme and inherited much of its pedagogy from this wider planned health intervention.

A group of such systems has been produced by the CARTE group who are strongly focused on what some might regard as training rather than education.

---

[3] We take pedagogy to be the principles and methods of instruction – see http://www.thefreedictionary.com/pedagogy

Nevertheless, the group is seriously concerned with pedagogical aspects of NILEs.

Johnson, Vilhjalmsson and Marsella's Tactical Language Training System (TLTS) is a system intended to help users learn a foreign language and become sensitive to cultural issues in a potentially dangerous situation. Learners are placed in a situation with a strong narrative theme – so the TLTS is another example of a NILE designed to teach some specific content. The learning of foreign languages in the TLTS system has a tightly defined purpose – oriented to the task of surviving in a potentially hostile environment [45]. The pedagogy is perhaps narrowly defined, being primarily trainer driven.

An interesting development in nursing education is worth mentioning here – primarily because the pedagogy is somewhat different from the pedagogies found in typical school settings. Diekelmann has introduced and advocated the use of *narrative pedagogy* within nursing education [74]. This approach has found application also within teacher education.

Narrative Pedagogy is placed somewhat in tension with standard approaches to learning in that the emphasis is on the generation of person-centred descriptions of situations and their interpretation within a community.

> "Teachers enacting Narrative Pedagogy challenge the conventional pedagogical assumptions that knowledge is the foundation for thinking, that thinking precedes action (behavior), and that action is evidence of a student's thinking, competence, and expertise. In conventional pedagogy, thinking is often reduced either to a process or an outcome akin to problem solving. To that end, teachers commonly specify learning outcomes in advance to ensure that students develop and demonstrate their thinking ability. ...Further embedded in this conventional approach are the assumptions that (1) the 'appropriate action' in a particular situation is clear and uncontested and can be readily identified by a student who is 'thinking'. To fail to determine the 'appropriate action' is evidence of a 'thinking problem'. And (2), the students who perform an 'appropriate action' did not do so by chance but by 'thinking' within the clinical situation." [75]

While Narrative Pedagogy might be seen as more relevant to teachers than to learners (though teachers are learners too), it stresses an aspect which is valuable to the designers of NILEs. The approach stresses the importance for the learner of dealing with situations through interpretation. It downplays the importance of being absolutely right or absolutely wrong, and of assessing learners through objective tests. It is similar to Narrative Inquiry in that argumentation is aimed at mutual understanding rather than winning [76]. For some, this will appear anathema (i.e. more or less taboo). For others taking a strong constructivist approach, it is not such an alien way of thinking about learning.

For school learners, the methods of Narrative Pedagogy may not always be appropriate, but it is close to some movements in education connected with inclusion, promotion of self esteem, recognising different kinds of individual achievement and so on. I would also argue that the underlying philosophy can be used as the theoretical grounding for future work on NILEs.

If NILEs can embody complex situations that encourage the learner to generate their own responses to the challenges found in a situation – whether this be in relation to understanding physics or how to respond to bullying – then we are part way to an approach that could support Narrative Pedagogy. What is evidently missing from most NILEs is the pooling of learner's interpretations and the opportunity for a learning community to work with these interpretations to form a new understanding.

4.5 The Next 10 Years

In the previous sections, we outlined some of the achievements of the last fifteen years. So what can be expected in, say, the next ten years? First, we consider the achievements in relation to the potential of NILEs outlined in the introduction.

The most impressive work on preconditions for learning seems to be that associated with the production and use of Ghostwriter and FearNot!. The kinds of learning may be quite different (story writing vs coping strategies for handling bullying) but both are very definitely about what happens after using the system. Of course, all learning systems share this, but these two NILEs are good examples of what can be achieved. We can expect more good examples of such systems. However, NILEs have mostly been used to set up the situations rather than provide bridges between what the learner knows and the intellectual and emotional challenges they are about to face. We would like to see more NILEs in the next ten that prepare learners in this manner.

In a similar manner, all learning systems seek to consolidate the learning in some manner. Ghostwriter, FearNot!, SAGA and the design process NCID all support the notion of reflection. SAGA highlights this aspect strongly, but perhaps future NILEs may provide stronger ways of organising the learner's experiences as part of reflection-on-action. SAGA provides for collaborative reflective experiences – we may hope for more such systems but perhaps with more extensive functionality for taking and transforming the stories of the learners.

Two exemplar systems (SAGA and FearNot!) concentrate on automating the management of learning. SAGA is particularly distinctive because of its use of reflection-on-action, and its emphasis of providing other systems with a 'plug in'. However, there are other NILEs of note introduced into our analysis – such as Carmen's Bright IDEAS – which demonstrate arguably more traditional management of learning functionality. We anticipate a small but growing number of such systems. However, these NILEs are very demanding of resources, and few research groups will be able to gather the necessary talents.

NCID provides a fascinating and potentially valuable variation on informant design. Other design frameworks are being used, and this has been a very fruitful area in the last few years. There is scope, however, for further work over the next few years. In particular, NCID has the potential for being effective with adult informants as well as children.

The evaluation area is still somewhat problematic in relation to NILEs – more problematic than the evaluation of environments designed to 'entertain'. The complexity of combining real learning gains within an entertaining environment is real, and some NILEs appear to succeed but the traditional experimental designs begin to be difficult to apply the more seriously we take the perspective (story) of the learner to be important. This is an area where researchers need to develop robust methods of investigation related to methods such as narrative inquiry.

As pointed out above the notion of narrative pedagogy as developed within nursing education raises some profound challenges. It would be good to see some NILEs inspired by such approaches to learning (and teaching).

4.6 The Way Forward

Should our efforts go into improving the capability of NILEs to motivate learners through managing experiences in a more effective manner? Should we look to building increasingly attractive believable agents so that through identification and empathy we might come to care about what they seem to care about (which, in our case, is connected with learning)? Should we work on the notion of narrative pedagogy in order that we harness the power of narrative for teaching?

In Shakespeare's Merchant of Venice, Bassanio is presented with a choice of three caskets made of gold, silver and lead. He has to choose the right casket to win Portia's hand in marriage. Are we in a similar situation? Is there a 'best' choice here, or can we choose to have all the caskets?

Having presented the three chests in which the answer to our quest might lie, which should we open? If our aim is improved learning, which 'option' is the one we should pursue? For example, is not the notion of a narrative pedagogy the worthy 'common drudge' of a goal we have actually been pursuing throughout this chapter? Is not the search for more attractive believable agents one that will produce self evident results – or is this 'gaudy' aim ultimately deceptive and disappointing? Is the 'real' winner the management of experiences?

My personal feeling is that all these aims are subservient to the one that makes sense of the others – the value of promoting good relationships between the learner, their teachers and system designers. In a previous paper, I pointed out that we needed to focus on building an empathic relationship with the learner [77]. Now I would like to suggest, in line with the notion of empathic design [64] that it becomes increasingly important to draw the system designer into a two-way relationship with every learner. This is not so different from Sims' perspective on design [78] which asks how to incorporate 'the reciprocal roles of both designer and user in the overall interactive process'.

> "... introduces the designer and user to the process, with both having the option to interact with the content and each other reciprocally. While in the interpersonal world this contact is clearly synchronous, a computer-based application also enables asynchronous contact. This allows the option for both synchronous and asynchronous mutual reciprocity (SMR and AMR) to be introduced to the interactive model." [78]

While education may be seen as functional, with an agenda to 'tick off' various achievements such as producing citizens with good communication skills and an acceptable ethical value system along with well developed literacy and numeracy skills, it can also be seen as a vehicle for the development of lifelong relationships with parents, teachers and peers – as well as employers. This emphasises the importance of personal relationships at a time when these are under stress in a world in which the impersonal, functional view of people seems to be dominant.

This is the hope for the future design for NILEs: that such systems will be of use in sustaining the personal development of learners in terms of building and supporting quality relationships with, amongst others, parents, teachers and fellow learners. In using NILEs of the future, we might hope that these system will also be used to help learners attain a wide range of competencies.

Appendix: Postscript

For those wanting to join in the search for NILEs, then look out for:

- The Narrative and Interactive Learning Environments series of conferences (e.g. NILE 2000, 2002, 2004, 2006) that are held every second year in Edinburgh, UK. See http://www.narrate.org for some information.
- The AI in Education conferences that occur every second year which have recently begun to feature a workshop on Narrative Environments – e.g. http://projects.ict.usc.edu/nle/ for AIED 2007 and http://gaips.inesc-id.pt/aied05-nle/ for the Narrative Learning Environments workshop at AIED 2005.
- The Interactive Conference on Virtual Storytelling series – see http://www.virtualstorytelling.com
- Special events organised by AAAI such as the AAAI 2007 Fall Symposium on Intelligent Narrative Technologies – see http://gel.msu.edu/aaai-fs07-int/. The influential 1999 Fall Symposium on Narrative Intelligence was published as a book [79].

- The AERA Narrative Inquiry SIG – email narrativesig@colostate.edu or visit http://www.narrativesig.cahs.colostate.edu.
- The UK EPSRC funded DAPPPLE (Drama and Performance for Pleasurable Personal Learning Environments) research network – see http://www.dappple.org.uk.
- The Special Interest Group on Narrative Environments supported by the EU funded Kaleidoscope network of excellence (http://nle.noe-kaleidoscope.org/). This group have collected some useful resources together. The book on technology-mediated narrative environments for learning is worth noting [80].
- Special issues of journals such as the International Journal of Continuing Engineering Education and Lifelong Learning 14(6) in 2004.

References

1. Mateas, M.: An oz-centric review of interactive drama and believable agents. In Wooldridge, M., Veloso, M., eds.: Artificial Intelligence Today. Springer, Berlin Heidelberg New York (1999) 297–328
2. Plowman, L., Luckin, R., Laurillard, D., Stratfold, M., Taylor, J.: Designing multimedia for learning: Narrative guidance and narrative construction. In: Proceedings of Computer-Human Interaction, CHI'99, Pittsburgh, PA (1999) 310–317
3. Waraich, A.: Telling stories – the role of narrative in intelligent tutoring systems. In: Proceedings of ED-MEDIA 98. AACE, Charlottesville, VA (1998)
4. Marsella, S.C., Johnson, W.L., LaBore, C.: Interactive pedagogical drama. In: Proceedings of the Fourth International Conference on Autonomous Agents. (2000) 301–308
5. Marsella, S., Johnson, W.L., LaBore, C.: Interactive pedagogical drama for health interventions. In Hoppe, H., Verdejo, M., Kay, J., eds.: Artificial Intelligence in Education: Shaping the Future of Learning Through Intelligent Technologies. IOS Press, Amsterdam (2003)
6. Weibelzahl, S.: Problems and pitfalls in the evaluation of adaptive systems. In Chen, S.Y., Magoulas, G.D., eds.: Adaptable and Adaptive Hypermedia System. IMR Press, Hershey, PA (2005) 285–299
7. Gagné, R.: The Conditions of Learning. 3rd edn. Holt-Saunders, New York (1977)
8. Raven, R.: The development of the concept of momentum in primary school children. Journal of Research in Science Teaching **5** (1968) 216–223
9. Ausubel, D., Novak, J., Hanesian, H.: Educational Psychology: A Cognitive View. Holt, Rinehart and Winston, New York (1978)
10. Romberg, T., Wilson, J.: The effect of an advance organizer, cognitive set, and post organizer on the learning and retention of written materials. Journal for Research in Mathematics Education **4**(2) (1973) 68–76
11. Keller, J.M.: Motivational design of instruction. In Reigeluth, C.M., ed.: Instructional-Design Theories and Models: An Overview of Their Current Status. Lawrence Erlbaum Associates, Hillsdale, NJ (1983)

12. Visser, J., Keller, J.M.: The clinical use of motivational messages: An inquiry into the validity of the ARCS model of motivational design. Instructional Science **19** (1990) 467–500
13. Barthes, R.: Introduction to the Structural Analysis of Narratives. In: Image Music Text. Fontana Press, London (1977)
14. Hall, R.H., Hall, C.R., Saling, C.B.: The effects of graphical post organization strategies on learning from knowledge maps. Journal of Experimental Education **67**(2) (1999) 101–112
15. Schon, D.A.: Educating the Reflective Practitioner. Jossey-Bass, San Francisco (1987)
16. Midoro, V., Chioccariello, A., Persico, D., Sarti, L., Tavella, M.: Ariadne's thread: An introduction to logic programming. Computers and Education **12**(1) (1988) 191–197
17. Waraich, A.: Designing Motivating Narratives for Interactive Learning Environments. PhD. thesis, Computer Based Learning Unit, Leeds University (2003)
18. Norman, D., Draper, S.: User Centered System Design. Lawrence Erlbaum Associates, Hillsdale, NJ (1986)
19. Scaife, M., Rogers, Y., Aldrich, F., Davies, M.: Designing for or designing with? Informant design for interactive learning environments. In: CHI'97: Proceedings of Human Factors in Computing Systems. ACM, New York (1997) 343–350
20. Chin, G.J., Rosson, M., Carroll, J.: Participatory analysis: Shared development requirements from scenarios. In Pemberton, S., ed.: Proceedings of CHI'97: Human Factors in Computing Systems. (1997) 162–169
21. Cooper, B., Brna, P.: Classroom conundrums: The use of a participant design methodology. Educational Technology and Society **3**(4) (2000) 85–100
22. Luckin, R., Underwood, J., du Boulay, B., Holmberg, J., Kerawalla, L., O'Connor, J., Smith, H., Tunley, H.: Designing educational systems fit for use: A case study in the application of human centred design for AIED. International Journal of Artificial Intelligence in Education **16**(4) (2006) 353–380
23. Clandinin, D.J., Connelly, F.M.: Narrative Inquiry: Experience and Story in Qualitative Research. Jossey-Bass, San Francisco (2000)
24. Robertson, J.: The effectiveness of a virtual role-play environment as a story preparation activity. PhD. thesis, Edinburgh University (2001)
25. Robertson, J., Oberlander, J.: Ghostwriter: Educational drama and presence in a virtual environment. Journal of Computer Mediated Communication **8**(1) (2002)
26. Robertson, J., Good, J.: Using a collaborative virtual role-play environment to foster characterisation in stories. Journal of Interactive Learning Research **14**(1) (2003) 5–29
27. Boström, A.: Chemistry narratives: How narratives from lived experience facilitate learning in chemistry narratives. In Brna, P., ed.: Proceedings of Narrative and Interactive Learning Environments: NILE 2006. (2006) 5–12
28. Brna, P.: Programmed rockets: An analysis of students' strategies. British Journal of Educational Technology **20**(1) (1989) 27–40
29. O'Shea, T.: Magnets, martians and microworlds: Learning with and learning by OOPS. In Bierman, D., Breuker, J., Sandberg, J., eds.: Artificial Intelligence and Education: Reflection and Synthesis. IOS, Amsterdam (1989) 193–194

30. Barab, S.A., Sadler, T.D., Heiselt, C., Hickey, D., Zuiker, S.: Relating narrative, inquiry, and inscriptions: Supporting consequential play. Journal of Science Education and Technology **16**(1) (2006) 59–82
31. Sharry, J., Brosnan, E., Fitzpatrick, C., Forbes, J., Mills, C., Collins, G.: 'Working things out' a therapeutic interactive cd-rom containing the stories of young people overcoming depression and other mental health problems. In Brna, P., ed.: Proceedings of Narrative and Interactive Learning Environments NILE 2004. (2004) 67–74
32. Smith, M.: Engaging Characters: Fiction, Emotion and the Cinema. Clarendon Press, Oxford (1995)
33. Paiva, A., Dias, J., Sobral, D., Aylett, R., Woods, S., Hall, L., Zoll, C.: Learning by feeling: Evoking empathy with synthetic characters. Applied Artificial Intelligence **19**(3–4) (2005) 235–266
34. Zoll, C., Enz, S., Schaub, H., Aylett, R., Paiva, A.: Fighting bullying with the help of autonomous agents in a virtual school environment. In Fum, D., Missier, F., Stocco, A., eds.: Proceedings of the 7th International Conference on Cognitive Modeling (ICCM 2006). Edizioni Goliardiche, Trieste (2006) 340–345
35. Weyhrauch, P.: Guiding Interactive Drama. PhD. thesis, School of Computer Science, Carnegie Mellon University (1997)
36. Laurel, B.: Computers as Theatre. Addison-Wesley, Reading (1993)
37. Sengers, P.: Schizophrenia and narrative in artificial agents. In Mateas, M., Sengers, P., eds.: Narrative Intelligence. John Benjamins, Amsterdam (2003) 259–278
38. Mateas, M., Stern, A.: A behavior language: Joint action and behavioral idioms. In Predinger, H., Ishiuka, M., eds.: Life-Like Characters: Tools, Affective Functions and Applications. Springer, Berlin Heidelberg New York (2004)
39. Vassileva, J.: Dynamic course generation on the www. In de Boulay, B., Mizoguchi, R., eds.: Artificial Intelligence in Education: Knowledge and Media in Learning Systems. IOS Press, Amsterdam (1997) 498–505
40. del Soldato, T.: Detecting and reacting to the learners motivational state. In Frasson, C., Gauthier, G., McCalla, G., eds.: Intelligent Tutoring Systems: Proceedings of the Second International Conference. Springer, Berlin Heidelberg New York (1992) 567–574
41. Murray, R.C., VanLehn, K., Mostow, J.: Looking ahead to select tutorial actions: A decision-theoretic approach. International Journal of Artificial Intelligence in Education **14**(3/4) (2004) 235–278
42. Szilas, N.: Stepping into the interactive drama. In Göbel, S., Spierling, U., Hoffmann, A., Iurgel, I., Schneider, O., Dechau, J., Feix, A., eds.: Technologies for Interactive Digital Storytelling and Entertainment: Second International Conference, TIDSE 2004. Springer, Berlin Heidelberg New York (2004)
43. Szilas, N.: IDtension: a narrative engine for interactive drama. In Göbel, S., Braun, N., Spierling, U., Dechau, J., Diener, H., eds.: Technologies for Interactive Digital Storytelling and Entertainment: TIDSE 03 Proceedings. Fraunhofer IRB, Darmstadt (2003)
44. Raybourn, E.M., Bos, N.: Design and evaluation challenges of serious games. In: CHI 2005. ACM (2005) 2049–2050
45. Johnson, W.L., Vilhjalmsson, H., Marsella, S.: Serious games for language learning: How much game, how much AI? In Looi, C.K., McCalla, G., Bredeweg, B., Breuker, J., eds.: Artificial Intelligence in Education: Supporting Learning

Through Intelligent and Socially Informed Technology. IOS Press, Amsterdam (2005) 306–313
46. Mateas, M., Stern, A.: Structuring content in the façade interactive drama architecture. In Young, R.M., Laird, J.E., eds.: Proceedings of the First Artificial Intelligence and Interactive Digital Entertainment Conference. AAAI Press, Stanford, CA (2005) 93–98
47. Murray, T.: Authoring intelligent tutoring systems: An analysis of the state of the art. International Journal of Artificial Intelligence in Education **10** (1999) 98–129
48. Riedl, M., Saretto, C., Young, R.: Managing interaction between users and agents in a multi-agent storytelling environment. In: Proceedings of the 2nd International Conference on Autonomous Agents and Multi-Agent Systems. (2003) 361–368
49. Young, R.M., Riedl, M.O.: Towards an architecture for intelligent control of narrative in interactive virtual worlds. In: Proceedings of the 2003 International Conference on Intelligent User Interfaces. (2003) 310–312
50. Riedl, M.O., Young, R.M.: A planning approach to story generation for history education. In Brna, P., ed.: Proceedings of Narrative and Interactive Learning Environments NILE 2004 (2004) 41–48
51. Machado, I.: Children, Stories and Dramatic Games: A Support and Guidance Architecture for Story Creation Activities. PhD. thesis, Computer Based Learning Unit, Leeds University (2004)
52. Machado, I., Brna, P., Paiva, A.: Tell me a story. Virtual Reality **9**(1) (2006) 34–48
53. Machado, I., Paiva, A., Brna, P.: Real characters in virtual stories: Promoting interactive story-creation activities. In Balet, O., Subsol, G., Torguet, P., eds.: Proceedings of ICVS 2001. Springer, Berlin Heidelberg New York (2001) 127–134
54. Machado, I., Brna, P., Paiva, A.: Learning by playing: Supporting and guiding story-creation activities. In Moore, J.D., Redfield, C.L., Johnson, W.L., eds.: Proceedings of the 10th International Conference on Artificial Intelligence in Education AI-ED 2001. IOS Press, Amsterdam (2001) 334–342
55. Prada, R., Machado, I., Paiva, A.: TEATRIX: Virtual environment for story creation. In Gauthier, G., Frasson, C., VanLehn, K., eds.: Proceedings of the 5th International Conference of Intelligent Tutoring Systems. Springer, Berlin Heidelberg New York (2000) 464–473
56. Heathcote, D.: Drama and learning. In Johnson, L., O'Neill, C., eds.: Collected Writing on Education and Drama. Northwestern University Press, Evanston, IL (1991) 90–102
57. Propp, V.: Morphology of the folktale. University of Texas Press, Austin (1968)
58. Greimas., A.J.: Structural Semantics: An Attempt at a Method. University of Nebraska Press, Lincoln (1983) Trans. Daniele MacDowell, Ronald Schleifer and Alan Veile
59. Machado, I., Brna, P., Paiva, A.: 1, 2, 3 . . . action! directing real actors and virtual characters. In Goebel, S., Spierling, U., Hoffmann, A., Iurgel, I., Schneider, O., Dechau, J., Feix, A., eds.: Technologies for Interactive Digital Storytelling and Entertainment, Second International Conference, TIDSE 2004. Springer, Berlin Heidelberg New York (2004) 36–41
60. Riedl, M., Lane, H., Hill, R., Swartout, W.: Automated story direction and intelligent tutoring: Towards a unifying architecture. In: AI and Education 2005

Workshop on Narrative Learning Environments, Amsterdam, The Netherlands, July 2005
61. Hall, L., Woods, S., Aylett, R.: Fearnot! involving children in the design of a virtual learning environment. International Journal of Artificial Intelligence in Education **16**(4) (2006) 327–351
62. Good, J., Robertson, J.: CARSS: A framework for learner-centred design with children. International Journal of Artificial Intelligence in Education **16**(4) (2006) 381–413
63. Waraich, A., Brna, P.: A narrative centred informant design approach for interactive learning environments. International Journal of Continuing Engineering Education and Life-Long Learning **18**(2) (2008)
64. Brna, P.: On the role of self esteem, empathy and narrative in the development of intelligent learning environments. In Pivec, M., ed.: Affective and Emotional Aspects of Human-Computer Interaction Game-Based and Innovative Learning Approaches. IOS Press, Amsterdam (2006) 237–245
65. Self, J.: The defining characteristics of intelligent tutoring systems research: ITSs care, precisely. International Journal of Artificial Intelligence in Education **10**(3–4) (1999) 350–364
66. Mallon, B., Webb, B.: Stand up and take your place: Identifying narrative elements in narrative adventure and role-play games. Computers in Entertainment **3**(1) (2005)
67. Marsh, T.: Towards Invisible Style of Computer-Mediated Activity: Transparency and Continuity. PhD. thesis, University of York, UK (2004)
68. Marsh, T.: Staying there: an activity-based approach to narrative design and evaluation as an antidote to virtual corpsing. In Riva, G., Davide, F., IJsselsteijn, W., eds.: Being There: Concepts, Effects and Measurement of User Presence in Synthetic Environments. IOS Press, Amsterdam (2003) 85–96
69. Wright, P.C., Monk, A.F.: The use of think-aloud evaluation methods in design. ACM SIGCHI Bulletin **23**(1) (1991) 55–57
70. Marsh, T.: Presence as experience: Film informing ways of staying there. Presence **12**(5) (2003) 538–549
71. Knickmeyer, R.L., Mateas, M.: Preliminary evaluation of the interactive drama façade. In: CHI 2005. ACM (2005)
72. Malone, T.: Towards a theory of intrinsically motivating instruction. Cognitive Science **5**(4) (1981) 333–369
73. Laaksolahti, J.: Methods for evaluating a dramatic game: Capturing subjective enjoyment of dramatic experiences. In Dettori, G., Gianetti, T., Paiva, A., Vaz, A., eds.: Technology-Mediated Narrative Environments for Learning. Sense, Rotterdam (2006) 123–132
74. Diekelmann, N.: Narrative Pedagogy: Heideggerian hermeneutical analyses of lived experiences of students, teachers, and clinicians. Advances in Nursing Science **23**(3) (2001) 53–71
75. Andrews, C.A., Ironside, P.M., Nosek, C., Sims, S.L., Swenson, M.M., C.Yeomans, K.Young, P., Diekelmann, N.: Enacting narrative pedagogy: The lived experiences of students and teachers. Nursing and Health Care Perspectives **22**(5) (2001) 252–263
76. Conle, C.: The rationality of narrative inquiry in research and professional development. European Journal of Teacher Education **24**(1) (2001) 21–33

77. Brna, P., Cooper, B., Razmerita, L.: Marching to the wrong distant drum: Pedagogic agents, emotion and student modelling. In: Proceedings of the 2nd International Workshop on Attitude, Personality and Emotions in User-Adapted Interaction, Sonthoven, Bavaria (2001)
78. Sims, R.: Interactivity or narrative? A critical analysis of their impact on interactive learning. In: Proceedings of ASCILITE'98, Wollongong, Australia (1998) 627–637
79. Mateas, M., Sengers, P.: Narrative Intelligence. John Benjamins, Amsterdam (2003)
80. Dettori, G., Giannetti, T., Paiva, A., Vaz, A.: Technology-Mediated Narrative Environments for Learning. Sense, Rotterdam, The Netherlands (2006)

5

Knowledge Acquisition for Configurable Products and Services

Alexander Felfernig

Intelligent Systems and Business Informatics, University Klagenfurt, A-9020 Klagenfurt, Austria, alexander.felfernig@uni-klu.ac.at;
http://www.configworks.com/Alexander.Felfernig

Summary. Mass Customization has been established as a new paradigm which is defined as the production of highly variant products and services under Mass Production pricing conditions. Knowledge-based configurator technologies are key-enabling technologies supporting the effective implementation of a Mass Customization strategy. In order to successfully deploy configurator applications in commercial environments, knowledge acquisition and maintenance concepts have to be provided which support knowledge engineers and domain experts in the effective implementation of knowledge bases and user interfaces. This chapter focuses on two innovative approaches which can significantly improve knowledge acquisition processes for configurable products and services.

5.1 Introduction

In the first part of the last century, Henry Ford produced the *T model* and revolutionized manufacturing processes by introducing Mass Production techniques (efficient production of a high number of identical products). The sentence *my customers can get every car color they want as long as it is black* perfectly represents the basic idea of Mass Production. Definitely, Mass Production of identical products is a business model of the past. Nowadays, buyer's markets predominate which impose new challenges to production and sales processes: companies are forced to produce products and services that meet the individual needs of a customer. In this context, the Mass Customization paradigm [1, 21] has been established which represents the customer individual production of highly variant (complex) products under (nearly) Mass Production pricing conditions. That means, the goal is not only to switch to a customer-individual production but as well to do this under time and pricing conditions of Mass Production. In this context, configuration [3, 11, 12, 17, 19, 20] and personalization technologies [2, 4, 8, 10, 14, 16] can be interpreted as major enabling software technologies for the implementation of a Mass Customization strategy. Configuration knowledge as well

A. Felfernig: *Knowledge Acquisition for Configurable Products and Services*, Studies in Computational Intelligence (SCI) **104**, 75–87 (2008)
www.springerlink.com

as personalization knowledge is stored in an organizational memory which includes formalized knowledge about product assortments and related marketing and sales strategies. Advantages of applying configuration technologies are, e.g., reduced error rates in the quotation phase since the feasibility of an offer is automatically checked by a configuration system. Personalization technologies help us to tackle the Mass-Confusion phenomenon [15], which means that customers are overburdened by the size and complexity of the offered product assortment. An effective application of both technologies can significantly improve customer satisfaction [7]. A major precondition for an effective application of those technologies is that configuration as well as personalization knowledge design and maintenance processes are accompanied by knowledge acquisition techniques supporting knowledge engineers and domain experts. Two such knowledge acquisition approaches are presented within the scope of this chapter.

The remainder of this chapter is organized as follows. First, the basic concepts of knowledge-based configuration are introduced (Sect. 5.2). In the following it will be shown how we can improve the effectiveness of configuration knowledge base development by the application of Software Engineering design languages (Sect. 5.3). In Sect. 5.4 an approach to the automated identification of faults in user interface descriptions of configurator applications is presented. The chapter is concluded with a short discussion of future research issues.

5.2 Knowledge-Based Configuration

What is configuration? The most well known definition of configuration is given by Sabin and Weigel [24]. They define configuration as a *special case of design activity were the artifact being configured is assembled from instances of a fixed set of well defined component types which can be composed conforming to a set of constraints.* The most famous rule-based configurator application has been developed by John McDermott in 1978 using the language OPS5: R1/XCON (Expert configurer) – a detailed discussion on the experiences with this system can be found in [3]. This configuration system has been developed for supporting sales processes related to DEC VAX computer systems. Typical versions of XCON included about 31,000 components and about 17,500 rules. Although XCON was quite successful, it had enormous maintenance problems which were a direct consequence of the intermingling of domain knowledge and related problem solving knowledge. If changes were needed in the domain knowledge, in many cases those changes influenced the order in which rules were fired and vice versa. This knowledge acquisition and maintenance challenge has been tackled by development of model-based approaches which are based on a strict separation of domain knowledge and problem solving knowledge. Most of today's commercially available configuration environments are based on a model-based

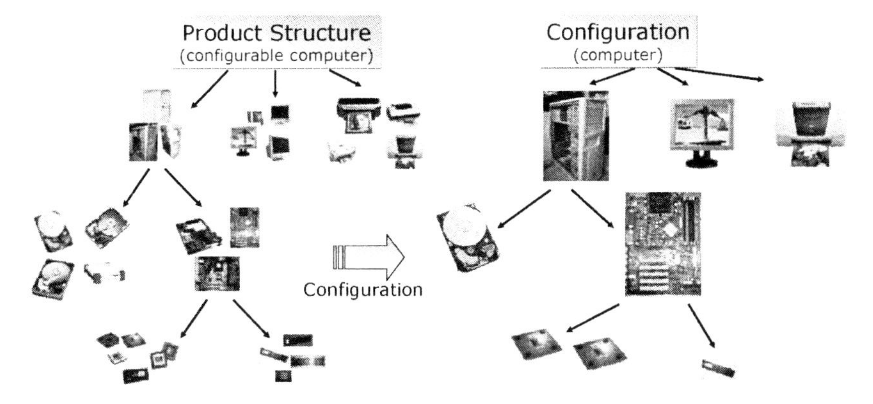

Fig. 5.1. Knowledge-based configuration

approach [11,13,17,19]; examples are the *configurator environments* of ORACLE, SAP, TACTON, and ILOG. Examples for *configurator applications* are financial services configurators developed for tasks such as portfolio configuration (abaXX, see, http://www.abaxx.de), car configurators such as Peugeot (http://www.peugeot.de) or BMW (http://www.bmw.at), or computer configurators – one of the most famous configurators in this context is the one provided by DELL (http://www.dell.com).

The following discussions in the chapter are based on a simple example for the configuration of computer systems (a sketch of the corresponding configuration model is depicted in Fig. 5.1). The needed ingredients for calculating a computer configuration for a concrete customer are the following:

- The definition of the product structure (in our case, the definition of the different components of a computer, e.g., motherboard, harddisk, cpu, etc.).
- A definition of a set of constraints which specify which components can be combined with each other (e.g., a motherboard of type A is *incompatible* with a CPU of type B).
- And finally, a specification of a set of customer requirements (also called key components [20]): which components does the customer require to be part of the final configuration?

Having specified product structure, constraints, and customer requirements, the configurator is able to calculate a solution (configuration) which is in the following presented to the customer (see Fig. 5.1). In our case, a configurator proposes a solution which includes a printer, a screen, one memory unit, a harddisk and two CPUs. This configuration result can in the following be assembled and delivered to a customer. On a more formal level, a configuration problem can be defined as follows:

Definition (Configuration Problem). *A configuration problem can be described as a logical theory describing a component library, related* *constraints, and system requirements (customer requirements). A configuration problem can defined by the triple (DD, SRS, CONL), where DD and SRS are sets of logical sentences and CONL is a set of predicate symbols which can be used for describing a concrete configuration. DD represents the domain description (component library and related constraints) and SRS specifies concrete requirements imposed by customers.*

In our case, a configuration CONF is described by ground literals conform to the predicate symbols of CONL which are of the form *type/2, conn/4*, and *val/3*, where *type(ID, t)* denotes the fact that the component *ID* is of type t, $conn(ID_1,\ p_1,\ ID_2,\ p_2)$ denotes the fact that component ID_1 is connected via port p_1 with component ID_2 via port p_2. An example for a concrete configuration is the following: type(ID_1, *cpua*). *type*(ID_2, *mba*). *conn*(ID_1, *mb*, ID_2, *cpu*). *va*$_l$(ID_1, *clockrate*, *500*). This simple configuration consists of two components (component ID_1 of type *cpua* and component ID_2 of type *mba*) which are connected via the ports *mb* and *cpu*. The value of the attribute *clockrate* of the component *cpua* has been set to *500* by the configuration system. Note that *ports* represent basic connection points between components part of a configuration result.

5.3 UML-Based Development of Configuration Knowledge Bases

The point now is that the development and maintenance of such formal representations (in our case logic-based) is a challenge for knowledge engineers as well as for domain experts. More intuitive representations are needed which allow us to reduce the knowledge acquisition bottleneck, provide more effective design support for knowledge engineers and domain experts and rapid prototyping processes which allow us to directly explore the behavior of a configuration knowledge base. The approach to improve this situation is to exploit widely known Software Engineering modeling languages (Unified Modeling Language (UML) [5, 23] and Object Constraint Language (OCL) [25]) for the design of configuration knowledge bases and to develop rules for translating those models into a corresponding formal (in our case logic-based) representation [6]. The process for the UML/OCL-based development of configuration knowledge bases is depicted in Fig. 5.2.

First, we have to define a configuration knowledge base on the basis of UML/OCL [6]. Using a UML profile for the development of configuration knowledge bases, we are able to check the syntactic correctness of the given configuration model. On the basis of a predefined set of translation rules, syntactically correct models are translated into a corresponding executable knowledge base. Such knowledge bases can directly be exploited for calculating concrete configurations. Before we can deploy a configuration knowledge base in a productive environment, we have to validate the calculated results. If this

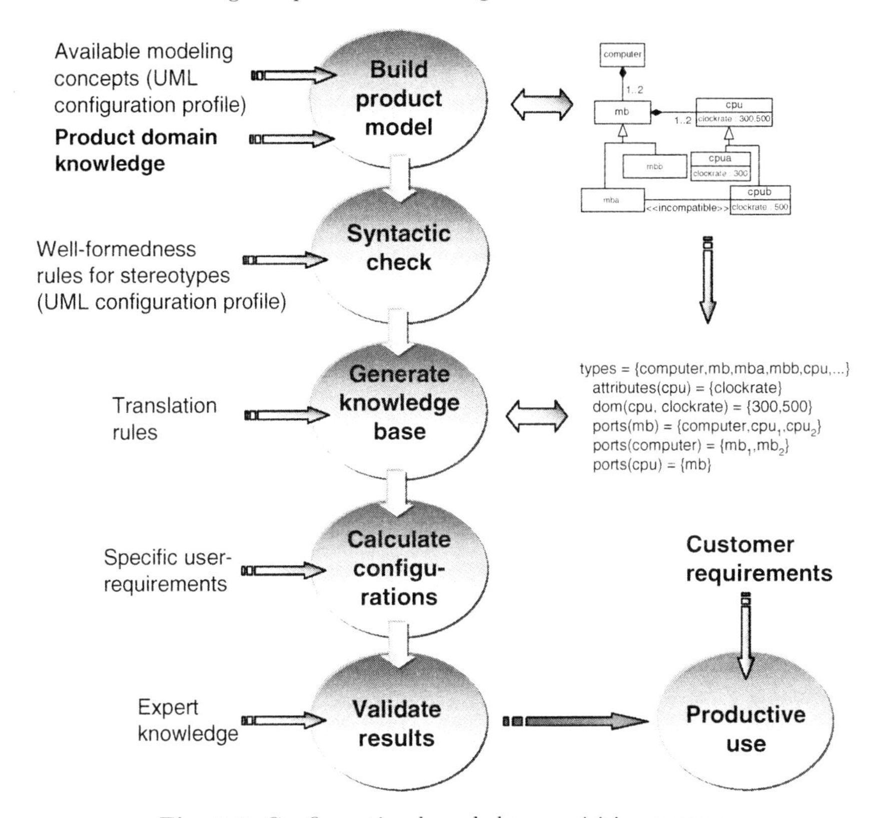

Fig. 5.2. Configuration knowledge acquisition process

validation process has been successfully completed, we are able to release the (maybe) new version of the knowledge base for productive use. The following example shows how UML-based configuration models are translated into a corresponding logic-based representation (see Fig. 5.3).

The central element of each configuration knowledge base is the product structure (see Fig. 5.3). In UML, such structures are defined in the form of part-of and generalization hierarchies, e.g., in our simple example, a configurable computer consists of one motherboard and each motherboard consists of at least one and at most two *cpus*. On the logical level, this structure can be represented as follows. Classes and their attributes are represented as component types with their attributes, e.g., a component type *cpu* has a corresponding attribute named *clockrate*. Furthermore, component types have defined connection points to remote component types, e.g., *each motherboard is connected with one or two cpus*. For generalization hierarchies we assume a *disjoint* and *complete* semantics which can be expressed as follows, e.g., *each motherboard is either a mba or a mbb*. A *mbb* is as well a *mb*. Aggregations are interpreted as being *composite*, e.g., *each cpu is connected to exactly one mb*.

```
DD (domain description) ⊇
{/* basic structure */
  types = {computer,mb,mba,mbb,cpu,...}.
  attributes(cpu)={clockrate}.
  dom(cpu, clockrate)={300,500}.
  ports(mb)={computer,cpu1,cpu2}.
  ports(computer)={mb}.
  ports(cpu)={mb}.

  /* generalization hierarchies */
  type(ID,mba) -> type(ID,mb).
  type(ID,mbb) -> type(ID,mb).
  type(ID,mb) -> type(ID,mba)
  type(ID,X) ∧ type(ID,Y) ∧ X ∈ {mba, mbb} -> Y ∈ {mb} ∨ X=Y.
  ...
  /* aggregations */
  type(ID,mb)  -> ∃(C) type(C,cpu) ^ conn(ID,CPU,C,mb) ^ CPU ∈ {cpu1,cpu2}.
  type(ID,cpu) -> ∃(M) type(M,mb) ^ conn(ID,mb,M,CPU) ^ CPU ∈ {cpu1,cpu2}.
  ...
}
```

Fig. 5.3. Translating UML-based product structures

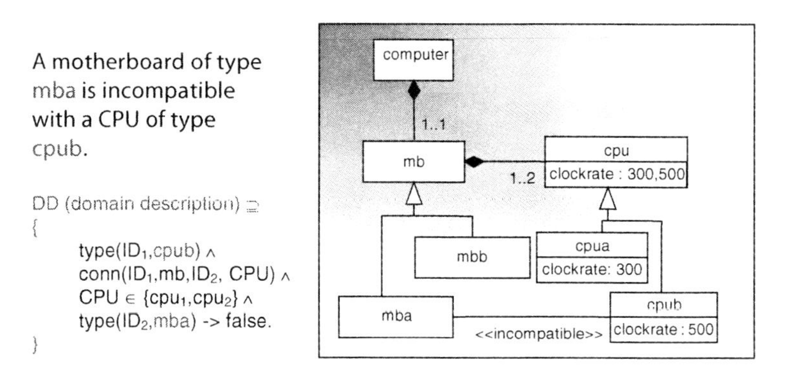

Fig. 5.4. Example UML constraint (incompatibility); the LHS depicts the corresponding logical representation

Having completed the derivation of the product structure, we have to translate additional constraints defined on the product structure. Figure 5.4 depicts a simple (graphical) UML constraint which denotes the fact that *a motherboard of type mba is incompatible with a cpu of type cpub*. For often used constraint types a corresponding graphical representation has been introduced, however, more complex constraints have to be directly specified in OCL [6].

A configuration result (configuration) is described by a set of components, attribute values and connections between components. For our example configuration knowledge base, a configuration result is the following: one computer connected to a motherboard, the motherboard is connected to one central processing unit (see Fig. 5.5).

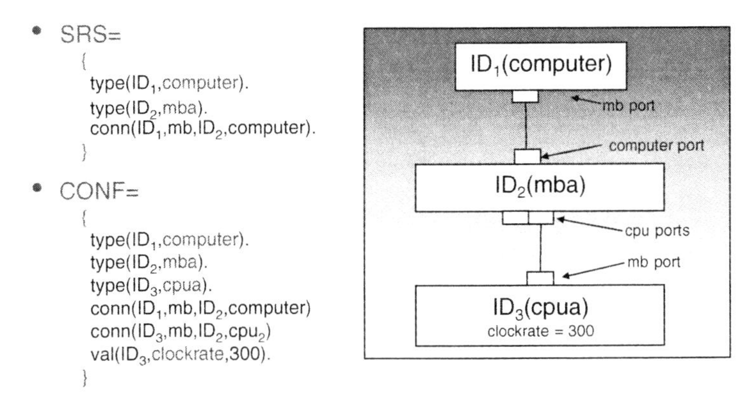

Fig. 5.5. Configuration result (CONF); SRS represent the corresponding customer requirements

The concepts presented here have been integrated into a corresponding knowledge acquisition environment – a Rational Rose Addin for the development of configuration knowledge bases (for details see [2, 6]). We now leave the topic of knowledge acquisition for configuration knowledge bases and take a look at a somewhat orthogonal aspect, the development of user interfaces for configurator applications. The goal here is to support knowledge engineers in the identification of faults in interface descriptions which are represented in the form of state charts. The approach is to apply concepts of model-based diagnosis [22] to identify faulty transition conditions in configurator user interface descriptions. The result of this work is an automated debugging environment for user interface descriptions [9].

5.4 Effective Development of Configurator User Interfaces

User interfaces for configurator applications can be represented in the form of specific types of state charts [9]. In our simple example (see Fig. 5.6) we have eight states (one initial state and one final state). State transitions are described by transition conditions, which specify the selection of consequent states depending on the given input of the user (answers to questions related to variables). In each state, users have to answer questions, depending on the answers a consequent state is selected. In our simple example (financial services configurator) the first question is related to the knowledge level of the customer. If this customer is an expert, then state q_1 is selected as consequent state.

When domain experts and knowledge engineers develop user interfaces for configurator applications, faulty transition conditions can be designed, where the following situations frequently occur (see Fig. 5.7):

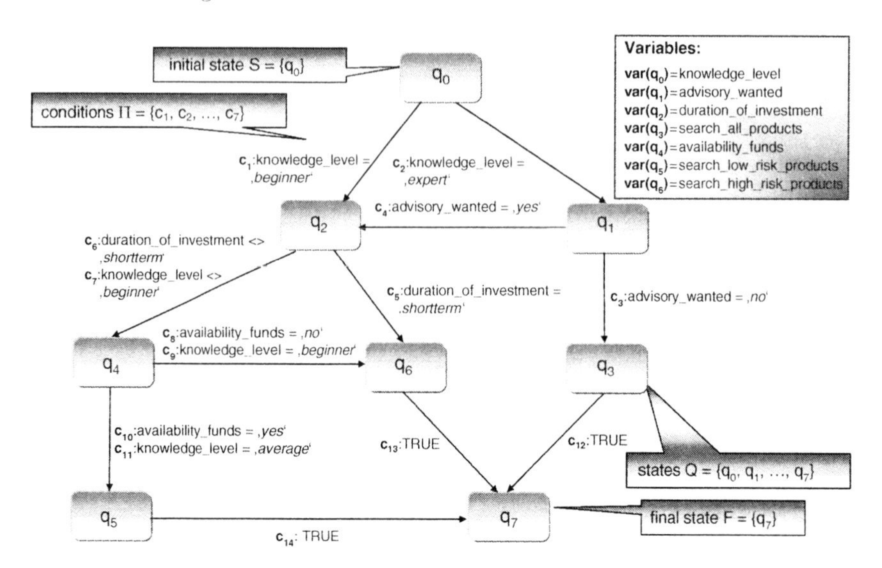

Fig. 5.6. Example configurator user interface description (user interface description of a financial services configurator)

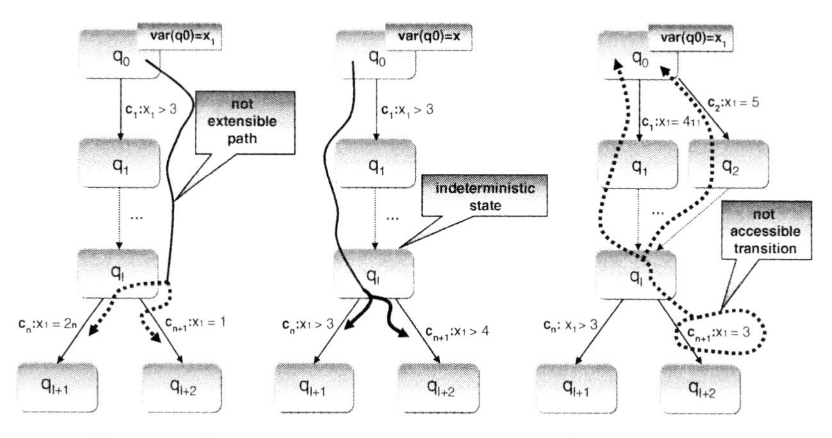

Fig. 5.7. Well-formedness rules for user interface descriptions

- *Not extensible paths*: here we have a situation where the path between the states q_0 and q_l is not extensible to any of the following conditions.
- *Indeterministic nodes*: here we have a situation where the state q_l allows path extensions to q_{l+1} as well as to q_{l+2}.
- *Not accessible transitions*: here we have a situation where one of the defined transition conditions is not accessible $(q_l,\ c_{n+1},\ q_{l+2})$, which definitely should not be the case.

From these examples we can derive a set of well-formedness rules which have to hold for interface descriptions. These rules are based on the definition of a *consistent path*.

Definition (Consistent Path). *Let* $p = [(q_1, C_1, q_2), (q_2, C_2, q_3), \ldots, (q_{i-1}, C_{i-1}, q_i)]$ *be a path from a state* q_1 *to a state* q_i. *p is consistent (consistent(p)) iff* $\cup C_j$ *is satisfiable. A consistent path p is extensible if there exists a transition condition* $(q_i,\ C_i,\ q_{i+1})$ *which is consistent with the transition conditions of p.*

Definition (Extensible Path). *Let* $p = [(q_1, C_1, q_2), (q_2, C_2, q_3), \ldots, (q_{i-1}, C_{i-1}, q_i)]$ *be a consistent path from a state* q_1 *to a state* q_i. *p is extensible (extensible(p)) iff* $\forall\ (q_i, C_i, q_{i+1}) : C_1 \cup C_2 \cup \ldots \cup C_{i-1} \cup C_i$ *is satisfiable.*

A transition condition $(q_i,\ C_i,\ q_{i+1})$ is *accessible* if there exists a path p, s.t. the transition conditions of p are consistent with $(q_i,\ C_i,\ q_{i+1})$.

Definition (Accessible Transition). *A transition* $t = (q_i, C_i, q_{i+1})$ *(postcondition of state* q_i*) is accessible (accessible(t)) iff there exists a path* $p = [(q_1, C_1, q_2),\ (q_2, C_2, q_3), \ldots, (q_{i-1}, C_{i-1}, q_i)] :\ C_1 \cup C_2 \cup \ldots \cup C_{i-1} \cup C_i$ *is satisfiable.*

In situations where the well-formedness rules are inconsistent with a given User Interface Description (see Fig. 5.8), i.e., the well-formedness rules WF_i should be consistent with the user interface description but are inconsistent, we have to identify a minimal set of transition conditions which have to be adapted in order to make the resulting interface description consistent with WF_i. Such a minimal set of faulty transition conditions can be determined by applying model-based diagnosis concepts [9]. The MBD approach starts with the description of a system which is in our case the structural description

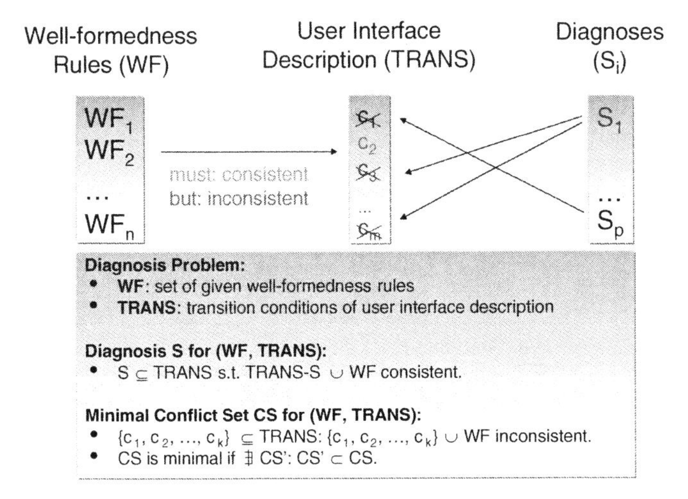

Fig. 5.8. Diagnosing faulty configurator user interface descriptions

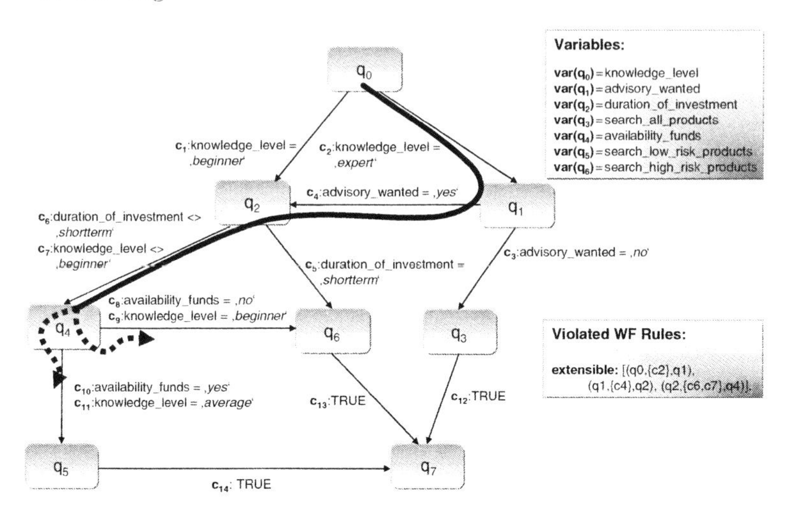

Fig. 5.9. Example: not extensible path

and the intended behavior of a configurator user interface (wellformedness rules). If the actual behavior conflicts with the intended system behavior, the diagnosis task is it to determine those system components (transition conditions) which, when assumed to functioning abnormally, will explain the discrepancy between the actual and the intended system behavior. Note that the calculated diagnoses need not to be unique, i.e., there can be different explanations for a faulty behavior.

In this context, a diagnosis problem is defined by a tuple (WF, TRANS), where WF represents the given set of well-formedness rules and TRANS represents the transition conditions of the user interface description. A diagnosis S for (WF, TRANS) is a set S subset of TRANS, s.t. TRANS-S ∪ WF consistent. A conflict Set CS for (WF, TRANS) is a subset of TRANS s.t. CS ∪ TRANS is consistent. CS is minimal if there does not exist CS' s.t. CS' is a proper subset of CS.

Our example user interface description is faulty. First, the path between q_0 and q_4 is not extensible (see Fig. 5.9). Second, there does not exist any path of which the transition conditions are consistent with the transition condition $(q_4, \{c_{10}, c_{11}\}, q_5)$ (see Fig. 5.10). A minimal diagnosis in this context is $\{c_1, c_{11}\}$, i.e., the transition conditions c_1, c_{11} have to be changed in order to make the user interface description consistent (a corresponding repair proposal is depicted in Fig. 5.10).

The algorithm for calculating diagnoses is an extended version of Reiters Hitting Set algorithm [22]. Diagnoses are generated in breadth-first manner – conflicts are provided by the QuickXPlain Algorithm proposed by [18]. There are two supported modes: all diagnoses calculated at once, and the interactive mode always calculating the next diagnosis. We have evaluated the performance of our algorithm using real-word user interface descriptions with representative erroneous transition conditions (see Fig. 5.11). The result of

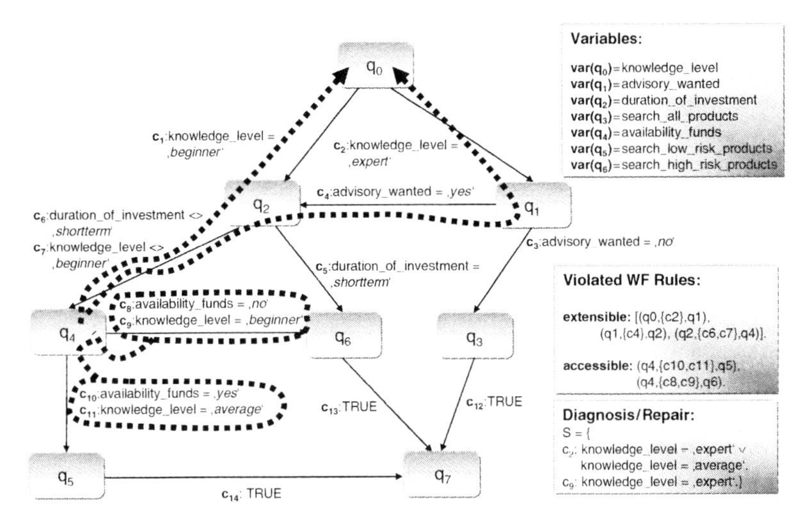

Fig. 5.10. Example: not accessible transition conditions

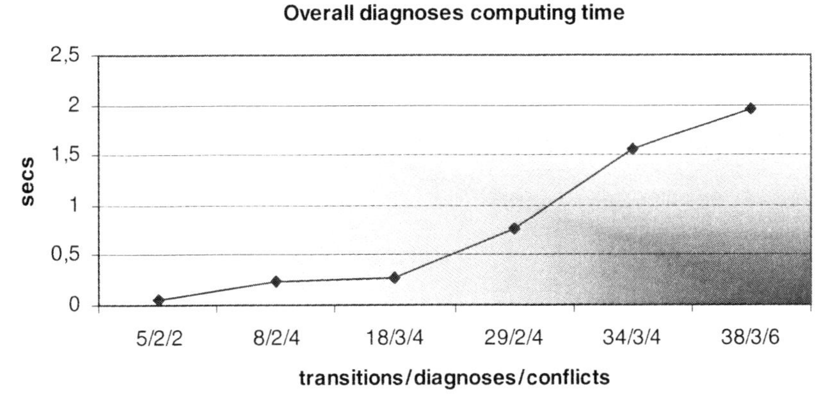

Fig. 5.11. Performance evaluation of diagnosis algorithm

this experiment shows the applicability of the presented diagnosis approach in interactive settings. For instance, descriptions with 38 transition conditions and six conflicts, and three corresponding diagnoses are calculated within the scope of two seconds.

The presented debugging concepts have been implemented for a commercially available configuration and personalization environment [9]. The performance of the presented modeling and debugging concepts has been extensively evaluated within the scope of commercial projects [8] as well as within the context of empirical studies (see, e.g., [7]). These studies show clear improvements in terms of time savings in development processes for user interfaces as well as in terms of reductions of faulty constraints introduced in user interface descriptions and knowledge bases. Our personalization environment [8,9] has been successfully exploited to deploy applications in domains such as investments, insurances, digital cameras, and e-government services. The debugging concepts are part of the test and debugging environment which has been implemented within the scope of the Koba4MS project [8,9].

5.5 Future Work

A major problem of current knowledge acquisition practices is that knowledge bases are developed by one expert (centralized approach). However, knowledge is distributed over different stakeholders. Knowledge acquisition environments should take into account this aspect: the future of knowledge acquisition will be environments supporting distributed design, testing and debugging processes. In addition to decentralized knowledge acquisition approaches, mechanisms for the determination of the optimal diagnosis are needed, e.g., by exploiting different complexity metrics from the area of Software Engineering for the determination of a knowledge bases complexity level. Finally, the construction of configuration knowledge bases can be significantly improved if we are able to analyze textual product descriptions, descriptions of sales processes and automatically derive initial versions of knowledge bases as well as process definitions. In this context, different approaches from natural language processing will be quite helpful.

5.6 Conclusions

This chapter shows how approaches from Software Engineering (UML/OCL) and Artificial Intelligence (Model-based Diagnosis) can be applied to improve the effectiveness of knowledge acquisition processes for configuration knowledge bases and configurator user interfaces. Such a support is crucial for the effective deployment of related applications in commercial environments.

References

1. Anderson, D.M. *Agile Product Development for Mass Customization.* McGraw-Hill, New York (1997)
2. Ardissono, L., Felfernig, A., Friedrich, G., Jannach, D., Petrone, G., Schaefer, R., and Zanker, M. A Framework for the development of personalized, distributed web-based configuration systems. *AI Magazine*, 24(3):93–108 (2003)
3. Barker, V.E., O'Connor, D.E., Bachant, J.D., and Soloway, E. Expert systems for configuration at Digital: XCON and beyond. *Communications of the ACM*, 32(3):298–318 (1989)
4. Burke, R. Knowledge-based recommender systems. *Encyclopedia of Library and Information Systems*, 69(32):180–200 (2000)

5. Dennis, A., Wixom, B., and Tegarden, D. *System Analysis and Design with UML Version 2.0: An Object Oriented Approach*, 2nd edn. Wiley, New York (2004)
6. Felfernig, A. Standardized configuration knowledge representations as technological foundation for mass customization. *IEEE Transactions on Engineering Management*, 54(1):41–56 (2007)
7. Felfernig, A., and Gula, B. An empirical study on consumer behavior in the interaction with knowledge-based recommender applications. *IEEE Conference on E-Commerce Technology (CEC06)*, pp. 288–296 (2006)
8. Felfernig, A., and Keiner, A. Knowledge-based interactive selling of financial services using FSAdvisor. *17th Innovative Applications of Artificial Intelligence Conference (IAAI'05)*, pp. 1475–1482, AAAI, Pittsburgh, PA (2005)
9. Felfernig, A., and Shchekotykhin, K. Debugging user interface descriptions of knowledge-based recommender applications. *ACM International Conference on Intelligent User Interfaces*, pp. 234–241, ACM, New York, Sydney, Australia, (2006)
10. Felfernig, A., Friedrich, G., and Schmidt-Thieme, L. Recommender Systems. *IEEE Intelligent Systems* (Special Issue on Recommender Systems), 22(3):18–21 (2007) IEEE
11. Fleischanderl, G., Friedrich, G., Haselboeck, A., Schreiner, H., and Stumptner, M. Configuring large systems using generative constraint satisfaction. *IEEE Intelligent Systems*, 13(4):59–68 (1998)
12. Forza C., and Salvador, F. Managing for variety in the order acquisition and fulfilment process: The contribution of product configuration systems. *International Journal of Production Economics*, (76):87–98 (2002)
13. Haag, A. Sales configuration in business processes. *IEEE Intelligent Systems*, 13(4):78–85 (1998)
14. Herlocker, J., Konstan, J., Terveen, L., and Riedl, J. Evaluating collaborative filtering recommender systems. *ACM Transactions on Information Systems*, 22(1):5–53 (2004)
15. Huffman, C., and Kahn, B. Variety for sale: Mass customization or mass confusion. *Journal of Retailing*, (74):491–513 (1998)
16. Jiang, B., Wang, W., and Benbasat, I. Multimedia-based interactive advising technology for online consumer decision support. *Communications of the ACM*, 48(9):93–98 (2005)
17. Juengst, E.W., and Heinrich, M. Using resource balancing to configure modular systems. *IEEE Intelligent Systems*, 13(4):50–58 (1998)
18. Junker, U. QUICKXPLAIN: Preferred explanations and relaxations for overconstrained problems. *19th National Conference on AI (AAAI04)*, pp. 167–172, AAAI, San Jose (2004)
19. Mailharro, D. A classification and constraint-based framework for configuration. *Artificial Intelligence for Engineering, Design, Analysis and Manufacturing Journal, Special Issue: Configuration Design*, 12(4):383–397 (1998)
20. Mittal, S., and Frayman, F. Towards a generic model of configuration tasks. *11th International Joint Conference on Artificial Intelligence*, pp. 1395–1401, Detroit, MI (1989)
21. Pine, B.J., and Davis, S. *Mass Customization: The New Frontier in Business Competition*. Harvard Business School Press, Boston, MA (1999)
22. Reiter, R. A theory of diagnosis from first principles. *AI Journal*, 23(1):57–95 (1987)

23. Rumbaugh, J., Jacobson, I., and Booch, G. *The Unified Modeling Language Reference Manual.* Addison-Wesley, Reading, MA (1998)
24. Sabin, D., and Weigel, R. Product configuration frameworks – A survey. *IEEE Intelligent Systems*, 13(4):42–49 (1998)
25. Warmer, J., and Kleppe, A. *The Object Constraint Language 2.0.* Addison-Wesley, Reading, MA (2003)

6

Interaction Modalities in Mobile Contexts

M.J. O'Grady[1], G.M.P. O'Hare[1], and S. Keegan[2]

[1] Adaptive Information Cluster (AIC), School of Computer Science and Informatics, University College Dublin (UCD), Belfield, Dublin 4, Ireland, michael.j.ogrady@ucd.ie, gregory.ohare@ucd.ie
[2] School of Computer Science and Informatics, University College Dublin (UCD), Belfield, Dublin 4, Ireland, stephen.keegan@ucd.ie

Summary. Enabling seamless and intuitive interaction is a long cherished objective of the HCI community. In classic desktop situations, the constituent processes have been studied over a long period of time and a mature understanding of the essential components has been obtained leading to broad agreement on best-practice principles and what constitutes good design. Though this endeavour has been of incalculable benefit, recent patterns of computer usage raise a new series of challenges that must be addressed. In particular, mobile computing is increasingly becoming the de facto usage paradigm: a situation that raises a new series of challenges for software engineers, and in particular, HCI professionals. In this chapter, the implications for one element of HCI in the mobile computing domain is examined, namely, interaction.

6.1 The Nature of the Mobile User

Mobile computer usage scenarios are intrinsically different from those associated with classic desktop computing. Mobile users may access services via their PDA or cell-phone in a diverse range of locations and circumstances. For example, they may require access in indoor or outdoor scenarios. Furthermore, they may be static, maybe standing on a footpath; or mobile, possibly travelling in a train. The prevailing situation may include significant ambient noise or inclement weather conditions.

Another key differentiator concerns the nature and capability of the average mobile device. It must be observed that such devices are notoriously resource-poor in contrast to standard desktop workstations. Likewise, the available telecommunications network may be characterised by high latency and low data-rates. All of which could potentially contribute to an unsatisfactory user experience. One redeeming characteristic concerns the nature of the mobile user's interaction. It is almost inherently short and to the point as the nature of average mobile usage scenarios, as well as the potential costs, do not lend themselves to users spending excessive time using mobile services.

M.J. O'Grady et al.: *Interaction Modalities in Mobile Contexts*, Studies in Computational Intelligence (SCI) **104**, 89–106 (2008)
www.springerlink.com

Gaming may be one notable exception. Before examining the various forms the mobile computing paradigm can assume, it is instructive to reflect on some historical developments in Human Computer Interaction (HCI).

6.1.1 Historical Developments in HCI

HCI [1,2] is inherently multidisciplinary. It has its roots in research that was conducted during the Second World War that sought to improve the usability and effectiveness of weapons. This resulted in the growth of ergonomics as a research discipline. In the 1960s, computing technologies had matured significantly, and problems were being identified that would ultimately lead to the establishment of the software engineering discipline. Implicit in some of these problems were interaction and usability issues.

In the 1980s, HCI became a discipline in its own right as the need for intuitive user interfaces and interaction techniques became apparent, due primarily to ongoing developments in personal computing. In the following decades, understanding of HCI issues would grow, leading to the present situation where HCI is perceived as a mature discipline in its own right, and one that forms an indispensable element of the software development process. It encompasses a rich suite of techniques, harvested from a range of disciplines including artificial intelligence [3], user modelling [4], personalisation [5], as well as various elements from the cognitive sciences [6,7].

As mobile computing becomes the predominant computer usage paradigm, the range of challenges facing HCI practitioners increases. Principles and heuristics that have been developed over time may not necessarily apply in mobile computing scenarios. Refinement and enhancement of acknowledged best practice principles may well be the preferred strategy, but its effectiveness will only become clear over time. However, the research community are aware of the challenges ahead, as the number of books [8,9], journal special issues [10,11] and conferences, especially MobileHCI [12], testifies. In the proceeding sections, the mobile computing paradigm is considered and the implications for interaction modalities are discussed.

6.1.2 Paradigms for Mobile Computing

Though most people would consider mobile computing as being exemplified by sophisticated mobile phones, or even traditional laptops, a number of visions of how mobile users may be best supported in the future have been proposed. It is probable that these may well be realised in specialised niche areas, at least initially:

- Ubiquitous computing [13] was conceived in the early 1990s and envisaged a world of artefacts augmented with embedded computational technologies, all linked transparently and seamlessly by a suite of networking technologies. Such artefacts would enable the provision of an array of

services to users, all of which could be accessed naturally and intuitively. In the intervening period, much research has taken place leading to significant developments in the prerequisite technologies, for example, Wireless Sensor Networks (WSNs).
- Wearable Computing [14] takes a diametrically opposite approach and envisages users possessing the required computational technologies about their very person, as distinct from being embedded in the fabric of the environment. Thus, they are not dependent on the environment in which they find themselves, and concerns about security and privacy are reduced considerably. Central to the wearable computing philosophy is the notion that computation is not in fact the user's primary concern or occupation. Rather, the wearable device should seek to aid the user in the fulfilment of their tasks, possibly through the use of Augmented Reality (AR) [15, 16].
- Ambient Intelligence (AmI) [17, 18] is a relatively recent initiative. It was conceived in response to the realization that as environments become increasingly saturated with computational technologies, people may find such environments hostile places to live and work, and therefore avoid them at any opportunity. Thus AmI promotes the engagement of Intelligent User Interfaces (IUIs) [19] as a means for minimizing the need for interaction, and improving the usability of the environment.

In each of these cases, it is noticeable that even though the hardware and software technologies employed all differ radically in their manifestation, the core objective in each case is identical: to make the mobile user's experience a satisfactory one. For the purposes of this discussion, the archetypical mobile user is considered to be one that is equipped with a PDA or, more likely, a mobile phone of similar specification. In the next section, interaction modalities in mobile computing scenarios are considered.

6.2 Interaction Techniques in Mobile Computing

QWERTY-style keyboards, augmented with a mouse, have proved the most popular method of interacting with personal computers for some time now. Alternative methods of interaction have been proposed and some have proved quite successful in niche domains. Examples include voice, handwriting, eye-gaze and gesture recognition systems. Of course these can be utilized in either a unimodal or multimodal fashion as the application demands. The advantage of such approaches is that they represent an inherently natural way of communicating and interacting. The difficulties with these, however, lies in the potential for error that exists during interpretation, as well as the computational resources required for implementing a software solution that can harness such modalities. This latter point is of critical importance and represents a significant obstacle when designing for mobile devices. Indeed, it has been demonstrated that the design of the Graphical User Interface (GUI) can affect the battery life of a mobile computer [20]. Thus, in the medium term at least, mobile HCI specialists are restricted in the techniques they can adopt during design and implementation.

As traditional interfaces, both from a hardware and stylistic perspective for example, the ubiquitous WIMP style, are impractical on standard mobile devices, alternative mechanisms of enabling physical interaction had to be developed. The net result has been the development of the key pads which most people are familiar with, and which have been standardized by the International Standards Organization [21]. In this layout, numeric keys are overloaded with alphabetic characters and other symbols; and though such a layout may seem counter intuitive, the success of the Short Message Service (SMS) suggests that significant numbers of people have adapted to, and mastered the use of, this layout.

As the power and sophistication of each succeeding generation of mobile devices increases, so does the demand for more sophisticated applications and services. In particular, the increase in screen real-estate to 1/4 VGA (320×240 pixels) has resulted in the possibility of deploying services that can incorporate a significant multimedia component. Thus, handset and operating system manufacturers have augmented their products with further support for interaction. Ironically, the result of this endeavour is an interface that partially resembles the WIMP style. Instead of a pointer, users can use a stylus. Alternatively, a navigation pad (5-way buttons) may be used. A soft keyboard is invariably supported by the Operating System (OS). Some manufacturers have actually included small physical keyboards, which are essentially stripped down QWERTY style with the function keys and numeric pad excluded.

Given the myriad of scenarios that the average mobile user may find themselves in, and the range of services that they may require at any moment in time, the interface designer is faced with some crucial decisions as to how best to design the interface for the application in question. It beholds designers to familiarize themselves with the potential application domain, the composition of the end-user base and, in particular, the nature of the task that must be performed. This will not guarantee the optimum design but will significantly increase the possibility of a satisfactory end-user experience. A discussion of the merits or otherwise of each interaction method is beyond the scope of this discussion. Rather, a more abstract yet holistic view of interaction will now be considered, in which the aforementioned interaction methods form singular examples.

6.2.1 Interaction as Intention

Though interaction can be considered in a number of ways, it is interesting to consider it in light of user intent. Broadly speaking, there are two extremes: one in which the user interacts consciously and explicitly with a software application; and at the other extreme, the user interacts unconsciously or

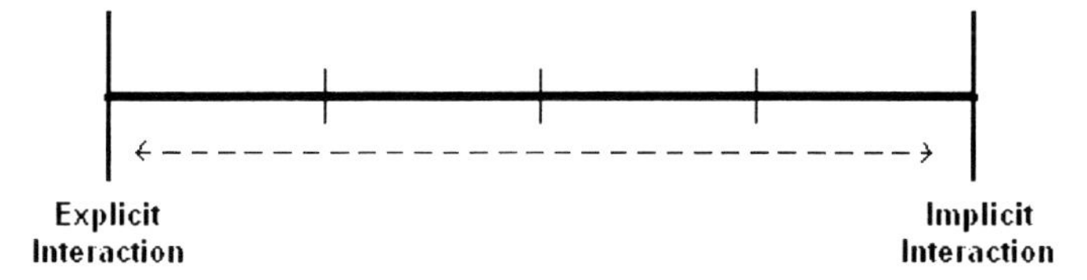

Fig. 6.1. The interaction continuum

implicitly (Fig. 6.1). In between, there are various degrees of each kind. However, any arbitrary interaction may encompass both an explicit and an implicit component to varying degrees.

Explicit Interaction

In this case, a user knowingly and purposely interacts with a software application. This may be accomplished by manipulating a GUI, running a command in a command window or issuing a voice command, for example. In short, the user intentionally performs some action, thus unleashing a series of events, resulting in an expected outcome. Most interaction with computers, both of the static and mobile variety, is explicit by nature.

Implicit Interaction

Implicit interaction [22] is, from a computation perspective, a more recent development and occurs when a user's subconscious actions (or lack of) indicate a preference that may be interpreted as an interaction. In itself, implicit interaction is not a new concept as people also communicate implicitly, for example by subconsciously smiling or frowning. As will be seen, to capture and interpret such interaction is difficult and subject to error. However, it offers useful possibilities in the mobile domain, assuming some necessary aspects of the user's behaviour can be identified.

As a practical example, consider the case of a museum visitor. By moving towards certain exhibits and viewing them for a significant time span could be interpreted by a mobile electronic tourist guide as an indication of a strong interest in the exhibit. Therefore the guide could deduce with a reasonably high degree of certainty that the visitor might welcome some additional information about that exhibit or might like to know of other similar exhibits in the museum.

Mixed Interaction

Any interaction may have an element of both the explicit and implicit modalities. For example, consider the case where a user encounters some difficulty with a software application. In some cases, but certainly not all, the application monitors the user activities. On noticing a pause that extends beyond a predefined threshold, the application may proactively intervene and ask the user if they need any assistance. Thus the users, by the length of time they take contemplating their next course of action, implicitly signal to the application that they might be experiencing some difficulty.

6.2.2 The Question of Context

Implicit interaction is closely associated with a user's context [23, 24] and some knowledge of the prevailing context is almost essential if designers want to incorporate it into their applications. In situations where the necessary elements of context can be quantified easily, the design and implementation process is straightforward. However, a potentially serious problem can arise when an incomplete model of the prevailing context is available; a situation that frequently occurs. Using techniques such as machine learning, Bayesian probability and so on, a confidence factor may be identified which will aid the decision process [25]. If the contextual situation still remains obscure, it may be necessary to initiate an explicit interaction with the user in question to clarify the situation. Ideally, this would not occur very often, but again, the nature of the application domain will determine how best to address this situation. Suffice to say that in a health monitoring application, the threshold value for initiating an explicit interaction would be lower than, for example, one that recommends local hostelries. However, determining a threshold and a subsequent course of action may involve significant computation, or possibly even a degree of intelligence.

Sophisticated computational techniques and algorithms almost invariable demand significant computational resources. Traditionally, this has made their deployment on mobile devices impractical; although with a high speed network, it is technically feasible to conduct the necessary processing on a fixed network node and return the results to the mobile client. However, ongoing developments in mobile hardware, operating systems and software toolkits have made the deployment of Artificial Intelligence (AI) techniques on such platforms feasible. One tractable solution from the field of Distributed AI [26] is the intelligent agent paradigm [27,28], and a number of toolkits that enable the deployment of such agents on mobile devices has been described in the literature [29] and are available via open-source. Before reflecting on the agent paradigm, it is necessary to briefly consider interaction from a software engineering viewpoint.

6.2.3 Interaction: A Software Engineering Perspective

Having established the necessity to include both explicit and implicit interaction modalities into the design of a mobile application, the task facing the software engineer is how to practically deliver this functionality. A useful classification is found in [30] where input modalities are categorised as either event based or streaming-based. Such a classification is derived from a software viewpoint, and as such, are not necessarily symmetric with human interaction modalities.

Event-Based Interaction Modalities

Practically all interaction is event-based and characterized by the production of discrete events. Examples include clicking a mouse, pressing a keyboard and so on. Discrete events are generated and interpreted by an appropriate software component. From a user perspective, event-based interaction is deliberate and indicates that the user is explicitly and consciously pursuing a course of action, with the expectation of a particular outcome.

Stream-Based Interaction Modalities

Stream-based interaction is characterised by the production of a stream of data, which must be captured and interpreted. A classic example of stream-based interaction would be voice-commands. In the case of implicit interaction, a stream of data from one or more data sources, for example sensors, must likewise be captured and analysed before an appropriate course of action can be identified. In contrast to the explicit interaction modality, however, deriving a meaningful interpretation of this data may require the availability of computationally-intensive software solutions. In addition, the mere process of continuously capturing the data streams may in itself have implications for software performance and ultimately usability, particularly in the case of mobile devices.

6.3 Intelligent Agents

Intelligent agents have been the subject of much research in recent years and applications of agents have been documented in numerous domains. Despite their popularity and appeal, the essential characteristics that constitute the term agenthood has been the subject of much debate. For the purposes of this discussion, agents will be considered as being either weak or strong [31]. Weak agents exhibit the following characteristics:

- Proactive: agents can initiate behaviours and courses of action that allows them fulfil their objectives.
- Reactive: agents can react to external events.
- Autonomous: agents can act independently of humans.
- Social: agents can communicate with other agents using an agreed Agent Communication Language (ACL) and ontology.

Strong agents incorporate the characteristics of weak agents but augment them with the following characteristics:

- Rationality: an agent will not act in such a way as would contradict its objectives.
- Benevolence: agents should not act in such as way as would harm another agent or compromise their host environment.
- Veracity: agents are truthful.

Depending on the nature of the application domain in question, it may not be necessary for an agent to incorporate all these characteristics. However, implicit in the notion of a strong agent is the availability of a sophisticated reasoning facility that can be used to reason about ongoing events and to plan accordingly.

In practice, any solution that is implemented using agents can also be implemented using conventional software approaches. However, it can be seen that agents encapsulate a number of characteristics that may appeal to software designers. However, there is no consensus as to when and where an agent-based solution should be used. In general, agents are perceived as offering alternative strategies for software development in areas that traditional techniques have not proved effective. Examples include domains that are complex and inherently dynamic.

6.3.1 Agent Architectures

Three broad categories of agent architecture have been identified:

1. Reactive agents act in a simple stimulus-response fashion and are characterized by a tight coupling between event perception and subsequent actions. Such agents may be modelled and implemented quite easily. The subsumption architecture [32], originating in research in behaviour-based robotics, is a classic example of a reactive architecture.
2. Deliberative agents can reason about their actions. Fundamental to such agents is the maintenance of a symbolic model of their environment. One popular implementation of the deliberative stance is the Belief-Desire-Intention (BDI) model [33], which is adopted by a number of well-documented agent frameworks, for example, Agent Factory [34] and JACK [35]. In the BDI scheme, agents maintain a model of their environment through a set of beliefs. Each agent has set of objectives or tasks that it seeks to fulfil, referred to as desires. By continuously monitoring its environment, an agent can detect opportunities when it is appropriate to carry out some of its desires. Such desires are formulated as intentions, which agent proceeds to realize.
3. Hybrid agents seek to harness the best aspects of each approach. A strategy that might be adopted would be to use the reactive component for handling discrete events, and the deliberative component for managing longer term objectives.

When considering an agent-based solution, it is necessary to identify a framework that both supports the required architecture, as well as the characteristics of the constituent agents.

6.3.2 Agent Characteristics and Interaction

On completing requirements analysis and, as part of the initial design stage, the software designer, possibly in conjunction with a HCI engineer, must decide on what modalities of interaction the proposed application will support. A number of factors must be considered including the nature of the task, the computational resources available as well as any common attributes of the end-user group. Should a decision be made in favour of supporting both the explicit and implicit interaction modalities, a software architecture that incorporates the necessary features to do this must be identified. A number of options exist, but from the previous discussion, it can be seen that certain characteristics of the agent paradigm are particularly suited for capturing interaction.

Explicit Interaction. Practically all applications must support this interaction modality. The reactive nature of agents ensures that they can handle this common scenario. As to whether it is prudent to use the computational overhead of intelligent agents just to capture explicit user input is debatable, particularly in a mobile computing scenario. Of course, if the application is modelled as a Multi-Agent System (MAS), then it is a logical course of action. However, the computational footprint of the agent should be estimated so that performance implications can be gauged.

Implicit Interaction. Utilizing the implicit interaction modality demands the continuous observation of the end-users' behaviour. As agents are autonomous, this does not present any particular difficulty, at least conceptually. However, there may be implications for performance, depending on the complexity of the application and the nature of the target mobile device. This situation is aggravated when it is considered that not only must some particular aspect of the user's behaviour be continuously monitored but the interpretation algorithms must likewise run in parallel. Thus there are implications here for the response time of the application, which, though it might unacceptable in an explicit interaction occurrence, may be adequate in an implicit interaction scenario. Quite how sophisticated the interpretation algorithms need be is domain dependent but the necessity for a deliberation component to undertake this interpretation process is obvious.

Recalling the need for an agent platform that operates on mobile devices, it is now appropriate to examine some frameworks that deliver this functionality.

6.3.3 Agent Systems for Mobile Devices

Deploying agents on mobile devices has, until recently, been unrealistic primarily due to hardware limitations. However, ongoing developments are increasingly rendering these limitations obsolete and a number of agent environments have been described in the literature, most of which are available under a General Public License (GPL). In some cases, existing platforms have been extended. For example, LEAP [36] has evolved from the JADE [37] platform. Likewise, microFIPA-OS [38] is an extension of the well-known open source platform FIPA-OS [39]. In the case of BDI agents, Agent Factory Micro Edition (AFME) [40] is one environment that supports such agents.

Historically, one classification of agents that is closely associated with user interfaces are the aptly named Interface Agents [41, 42]. Maes [43] has done pioneering work in this area and regards interface agents as potential collaborators with users in their everyday work and to whom certain tasks could be delegated. Ideally, interface agents would take the form of conversational characters which would interact with the user in a social manner. Though the focus of this research has been on classic workstation interfaces, there is no reason why such agents cannot be used in mobile computing scenarios. In particular, recent mobile devices can support rich multimedia applications and the necessary toolkits to develop such applications are becoming available. In the case of using agents as a basis for mobile HCI, some researchers have reflected on this [44, 45], but much more can be anticipated as the restrictions imposed by hardware limitations ease.

Before concluding this discussion, it is useful to reflect on some documented systems that demonstrate the potential use of intelligent techniques for mobile users. For example, Satoh [46] describes a framework, based on mobile intelligent agents that enable personalized access to services for mobile users. Grill et al. [47] describe an environment that supports the transfer of agent functionality into everyday objects. In the tourist domain, two mobile tourist guides, Crumpet [48] and Gulliver's Genie [49, 50] have been designed around the agent paradigm. Finally, Hagras and his co-researchers describe iDorm [51], an intelligent dormitory that uses embedded agents to realize an AmI environment. Techniques based on fuzzy logic are used to derive models of user behaviours. There are many more applications described on the WWW and elsewhere, and the interested reader is encouraged to consult the academic literature on Artificial Intelligence, Knowledge-based Systems, Intelligent information Systems and related topics. Conferences in these areas frequently have dedicated tracks covering ongoing developments. Usually, the proceedings are published online and stored in digital repositories where the documents may be searched by keyword. Two exemplar digital libraries are maintained by the Association for Computing Machinery (ACM) [52] and the Institute of Electrical & Electronic Engineers (IEEE) [53]. However, practically all the commercial publishing houses now maintain digital libraries; though a subscription is usually required for access.

6.4 Case Study: EasiShop

Having reflected on the various interaction modalities and motivated the need for an agent-based solution, the design and implementation of a solution that harnesses the use of agents for interaction can be considered. EasiShop [54, 55] is a functioning prototype mobile computing application, developed to illustrate the validity of the mobile-commerce (m-commerce) paradigm. By augmenting m-commerce with intelligent and autonomous components, the significant benefits of convenience and added value may be realized for the average shopper as they wander their local shopping mall or high street. In the remainder of this section, the synergy between agents and interaction is demonstrated through an illustration of an archetypical EasiShop usage scenario.

6.4.1 Interaction in EasiShop

Key steps in the modus operandi of EasiShop may be seen in Fig. 6.2. From an interaction perspective there are three key phases of interaction in EasiShop.

Construction of Shopper Profile and Shopping List (Explicit Interaction)

EasiShop is initiated when the shopper constructs a seed user profile. This comprises pertinent aspects of the shopper's personal profile such as age and gender. A further set of generalized information, alluding to the type of product classes which are of interest, is then obtained from the shopper. This latter information is obtained when the shopper constructs a shopping list. The list includes details of what products are sought and, to a certain extent, under what terms acquisition of these products is permissible.

Shopper Behaviour (Implicit Interaction)

Once the shopper has specified their profile information and has constructed a shopping list (Fig. 6.3a), the various components collaborate in a transparent and unobtrusive manner to facilitate the acquirements of all items on this shopping list. To manage this process, a certain degree of coordination is required, hence the need for sociability and mobility.

As the shopper wanders their local high-street, a proxy agent migrates transparently from their device into a proximal store. From here, the agent may migrate to an open marketplace where representative agents from a number of shops in the locality (including the current proximal store) may vie for the shopper's agent's custom. This process entails a reverse auction whereby the shopper's list is presented to the marketplace. Interested parties request to enter the ensuing auction and a set of the most appropriate selling candidates

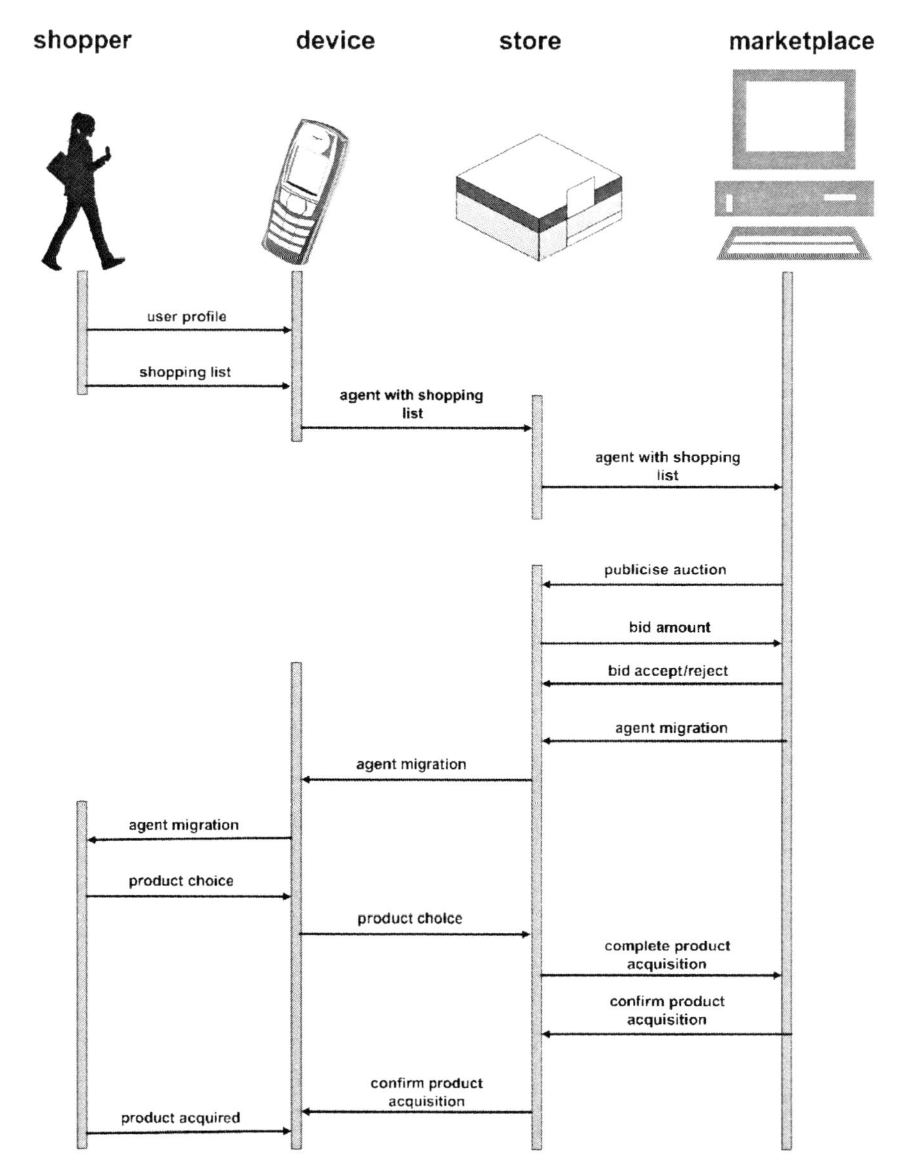

Fig. 6.2. Key steps in an EasiShop usage scenario

is chosen by the shopper's agent upon completion of the auction (if any). At this point the shopper's agent is ready to return to the shopper's device and will attempt to do so, though it may need to return from a different network access point.

Fig. 6.3. Shoppers must explicitly inform EasiShop about their requirements (**a**). EasiShop implicitly monitors the shopper behaviour and autonomously negotiates with nearby shops on their behalf (**b**)

Buying Decision (Explicit Interaction)

When the agent has returned to the shopper's device, it is laden with a set of product offerings resulting from the auction process. This set is presented to the shopper from whom an explicit decision is requested as to which product (if any) is the most acceptable (Fig. 6.3b). Once this indication has been made, the shopper is free to collect the item from the relevant shop, after completing the payment transaction. This decision is used as reinforcement feedback in that selection data is garnered to determine what kind of choice is likely in the future. Should the result of the auction not meet the agent's initial requirements, the shopper is not informed of what has taken place. However, the process will be repeated until such time as the shopping list is empty.

6.4.2 Architecture of EasiShop

EasiShop is modelled on a three-tiered distributed architecture (Fig. 6.4). From a centre-out perspective, the first tier is a special centralized server called the Marketplace. The design of the Marketplace permits trade (in the form of reverse auctions) to occur. The second tier is termed the Hotspot. This is a hardware and software hybrid suite, situated at each participating retail outlet, which is permanently connected to the Marketplace, and which allows the process of moving (migrating) representative (selling) agents from the retailers together with (buying) agents representing shoppers to the Mar-

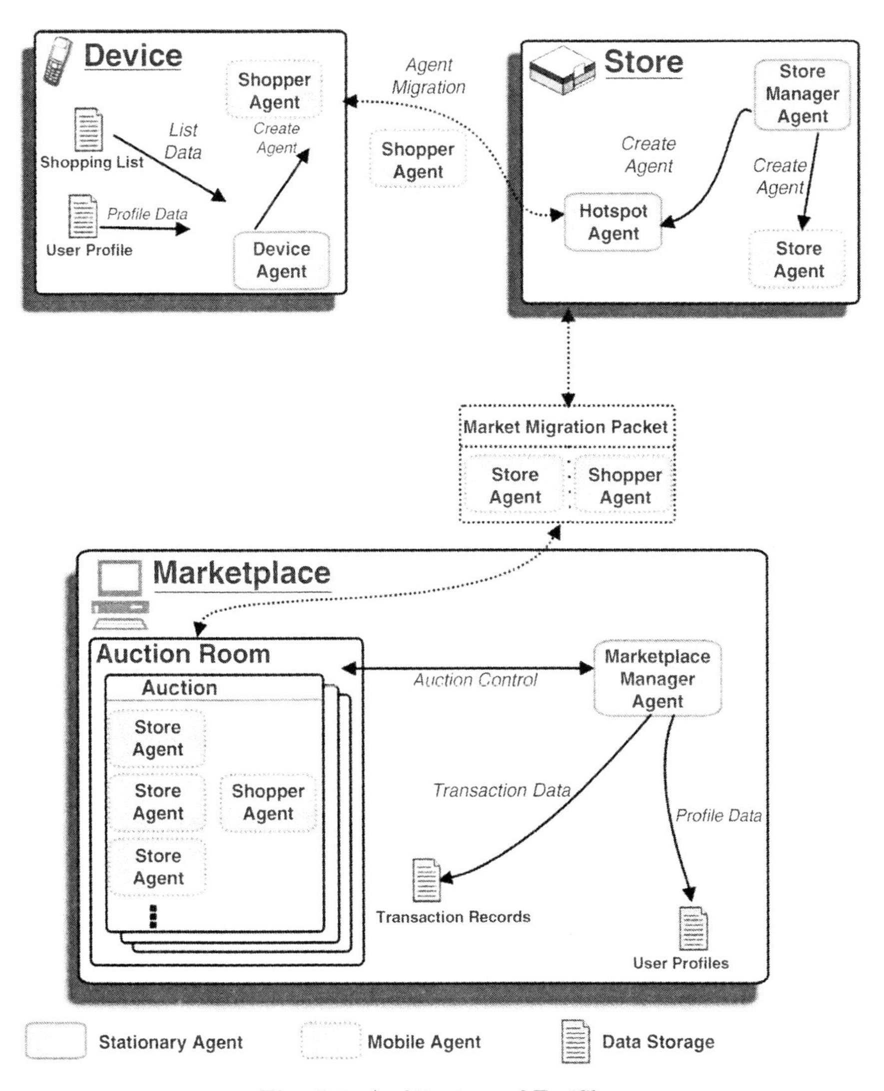

Fig. 6.4. Architecture of EasiShop

ketplace. The final and outermost tier in this schema is collection of device nodes, usually of the smart mobile phone or PDA genre.

6.4.3 Utilization of Mobile Intelligent Agents

The principal challenge when devising appropriate agents in a preference-based auctioneering domain is to deliver *personality embodiment*. These personalities represent traits of buyers (shoppers) and sellers (retailers) and are required to encapsulate *dynamism* (in that traits change over time) and *persistency* (in that a record of the traits needs to be accessible to the agents). These requirements are satisfied by the utilization of file-based XML data storage which contains sets of values upon which traits depend. For example, the characteristics of a store are encapsulated in an XML file containing rules which can formalize attributes like preferences for certain types of shopper (of a certain age or gender), temporal constraints (like pricing products according to time of day) and stock restrictions (like pricing products according to current stock levels).

Autonomy and mobility are essential to realizing implicit interaction in EasiShop. As the shopper explores their local high street, their agents continuously monitor their behaviour. On identifying that the shopper is near a shop, the agent autonomously migrates to a host server in the store, negotiates with the agents there, and if necessary, commences an auction with all interested nearby stores. In this way, the best deal is obtained for the shopper, at least within the negotiating constraints. All this activity takes place quickly and transparently while the shopper is within the vicinity of the shop. Only when there is a need to verify a purchase does the shopper become aware that their agents, influenced by their movements, have been acting on their behalf. From a hardware perspective, the mechanism to deliver agent mobility is implemented using a short-range communications protocol, namely Bluetooth.

6.5 Conclusion

As mobile devices increase in power and sophistication, the range and complexity of applications that they can support will likewise increase. However, the core mobile context is inherently different from the static context that most people are familiar with, and provides a radically different range of challenges to software designers. In parallel, opportunities may open up that never arise in static contexts and the availing of these opportunities offers exciting possibilities for improving and augmenting the mobile user's experience. Implicit interaction is one obvious example. However, assigning semantic meaning to the data streams that implicit interaction invariably produces calls for the engagement of sophisticated AI techniques. Though limited by mobile hardware and software technologies, the adoption of such techniques has become increasingly practical. In particular, the intelligent agent paradigm is one that has become computationally tractable on resource-limited devices and offers designers a viable option for incorporating AI techniques into their mobile applications and services as the need arises.

Acknowledgements

This material is based upon works supported by the Science Foundation Ireland (SFI) under Grant No. 03/IN.3/1361.

References

1. Dix, A., Finley, J., Abowd, G., Beale, R.: Human–Computer Interaction. 3rd edn. Prentice-Hall (2004)
2. Helander, M.G., Landauer, T.K., Prabhu, P., Prabhu, P.V. (eds): Handbook of Human–Computer Interaction. 2nd edn. Elsevier Science (1997)
3. Huang, Th.S., Nijholt, A., Pantic, M., Pentland, A. (eds): Artificial Intelligence for Human Computing, Lecture Notes in Artificial Intelligence, Vol. 4451. Springer (2007)
4. Fischer, G.: User Modeling in human–computer interaction. User Modeling and User-Adapted Interaction 11(1–2) (2001) 65–86
5. Kramer, J., Noronha, S., Vergo, J.: A user-centered design approach to personalization. Communications of the ACM 43(8) (2000) 44–48
6. Hollan, J., Hutchins, E., Kirsh, D: Distributed cognition: toward a new foundation for human–computer interaction research. ACM Transaction on Computer–Human Interaction 7(2) (2000) 174–196
7. Gray, W.D., Young, R.M., Kirschenbaum, S.S.: Introduction to a special issue on cognitive architectures and human–computer interaction. Human Computer Interaction 12 (1997) 301–309
8. Lumsden, J. (ed): Handbook of Research on User Interface Design and Evaluation for Mobile Technology, Information Science Reference (2008)
9. Love, S.: Understanding Mobile Human–Computer. Butterworth-Heinemann (2005)
10. Chittaro L. (ed): Special issue on HCI aspects of mobile devices and services. Personal and Ubiquitous Computing Journal 8(2) (2004)
11. Paterno, F.: Understanding interaction with mobile devices. Interacting with Computers, 15(4) (2003) 473–478
12. Human Computer Interaction with Mobile Devices and Services (MobileHCI) – http://www.mobilehci.org
13. Weiser, M.: The Computer for the Twenty-First Century. Scientific American. (1991) 94–100
14. Rhodes, B.J., Minar, N., Weaver, J.: Wearable computing meets ubiquitous computing: Reaping the best of both world. In: Proceedings of the Third International Symposium on Wearable Computers, IEEE Computer Society (1999) 141–149
15. Billinghurst, M. and Kato, H.: Collaborative augmented reality. Communications of the ACM 45(7) (2002) 64–70

16. Azuma, R., Baillot, Y., Behringer, R., Feiner, S., Julier, S., MacIntyre, B.: Recent advances in augmented reality. IEEE Computer Graphics and Application 21(6) (2001) 34–47
17. Aarts, E, Marzano, S. (eds): The New Everyday: Views on Ambient Intelligence. 010 Publishers, Rotterdam (2003)
18. Vasilakos, A., Pedrycz, W (eds): Ambient Intelligence, Wireless Networking, Ubiquitous Computing, Artech House, Norwood (2006)
19. Langley, P.: Machine learning for adaptive user interfaces. In: Brewka, G., Habel, C., Nebel, B. (eds): Advances in Artificial intelligence, Lecture Notes in Computer Science, Vol. 1303. Springer (1997) 53–62
20. Vallerio, K.S., Zhong, L., Jha, N.K.: Energy-efficient graphical user interface design. IEEE Transactions on Mobile Computing 5(7) (2006) 846–859
21. ISO/IEC 9995-8: Information technology – Keyboard layouts for text and office systems – Part 8: Allocation of letters to the keys of a numeric keypad. International Organization for Standardization (1994)
22. Schmidt, A.: Implicit Human Computer Interaction through Context. Personal Technologies 4(2,3) (1999) 191–199
23. Tamminen, S., Oulasvirta, A., Toiskallio, K., Kankainen, A.: Understanding mobile contexts. Personal and Ubiquitous Computing 8(2) (2004) 135–143
24. Dey, A.K.: Understanding and using context. Personal and Ubiquitous Computing 5(1) (2001) 4–7
25. Ranganathan, A., Al-Muhtadi, J., Campbell, R.H.: Reasoning about uncertain contexts in pervasive computing environments. IEEE Pervasive Computing 3(2) (2004) 62–70
26. O'Hare, G.M.P., Jennings, N.R. (eds): Foundations of Distributed Artificial Intelligence. Wiley (1996)
27. Wooldridge, M. J.: Introduction to Multiagent Systems. Wiley (2001)
28. Nwana H.: Software agents: An overview. Knowledge Engineering Review 11(3) (1996) 205–244
29. O'Hare, G.M.P., O'Grady, M.J., Muldoon, C., Bradley, J.F.: Embedded agents: A paradigm for mobile services. International Journal of Web and Grid Services 2(4) (2006) 379–405
30. Obrenovic, Z., Starcevic, D.: Modeling multimodal human–computer interaction. IEEE Computer 37(9) (2004) 65–72
31. Wooldridge, M., Jennings, N.R.: Intelligent agents: Theory and practice. The Knowledge Engineering Review 10(2) (1995) 115–152
32. Brooks, RA.: Intelligence without representation. Artificial Intelligence 47 (1991) 139–159
33. Rao, A.S., Georgeff, M.P.: BDI Agents: From theory to practice. In: Proceedings of the First International Conference on Multi-Agent Systems (ICMAS'95). AAAI Press (1995) 312–319
34. G.M.P. O'Hare: Agent factory: An environment for the fabrication of distributed artificial systems. In: O'Hare, Jennings (eds): Foundations of Distributed Artificial Intelligence, Wiley (1996) 449–484
35. JACK – The Agent Oriented Software Group, http://www.agent-software.com
36. Adorni, G., Bergenti, F., Poggi, A., Rimassa, G.: Enabling FIPA agents on small devices. In: Klusch, M., Zambonelli, F. (eds): Cooperative Information Agents V. Lecture Notes in Computer Science Vol. 2182. Springer (2001) 248–257
37. Bellifemine, F., Caire, G., Greenwood, D.: Developing Multi-Agent Systems with JADE. Wiley (2007)

38. Tarkoma, S., Laukkanen, M. (2002) Supporting software agents on small devices. In: Proceedings of the First International Joint Conference on Autonomous Agents and Multiagent Systems: Part 2 (AAMAS'02). ACM Press (2002) 565–566
39. Foundation for Intelligent Physical Agents (FIPA), http://www.fipa.org
40. Muldoon, C., O'Hare, G.M.P., Collier, R.W., O'Grady, M.J.: Agent factory micro-edition: A framework for ambient applications. In: Alexandrov, V., Van Albada, G., Sloot, P., Dongarra, J. (eds): Computational Science III. Lecture Notes in Computer Science, Vol. 3993 (2006) 727–734
41. Lieberman, H. 1997. Autonomous interface agents. In: Pemberton, S. (ed): Proceedings of the SIGCHI Conference on Human Factors in Computing Systems. ACM Press New York (1997) 67–74
42. Lieberman. H.: Letizia: An agent that assists web browsing. In: Mellish, C. (ed): Proceedings of the Fourteenth International Joint Conference on Artificial Intelligence (IJCAI-95). Morgan Kaufmann (1995) 924–929
43. Maes, P.: Agents that reduce work and information overload. Communications of the ACM 37(7) (1994) 30–40
44. Sas, C., O'Grady, M.J., O'Hare, G.M.P.: Electronic navigation – Some design issues. In: Chittaro, L. (ed): Lecture Notes in Computer Science (LNCS), Vol. 2795, (2003) 471–475
45. O'Hare, G.M.P., O'Grady, M.J.: Addressing mobile HCI needs through agents. In: Paterno. F. (ed): Lecture Notes in Computer Science (LNCS), Vol. 2411 (2002) 311–314
46. Satoh, I.: Software agents for ambient intelligence. In: Proceedings of IEEE International Conference on Systems, Man and Cybernetics (SMC'2004). IEEE Computer Society (2004) 1147–1150
47. Grill, T., Ibrahim, I.K., Kotsis, G.: Agents visualization in smart environments. In: Proceedings of the 2nd International Conference on Mobile Multimedia (MOMM2004). Oesterreichische Computer Gesellschaft (2004) 361–370
48. Poslad, S., Laamanen H., Malaka, R., Nick, A., Zipf, A.: Crumpet: Creation of user-friendly mobile services personalized for tourism. In: Proceeding of the Second IEE International Conference on 3G Mobile Communication Technologies. IEE (2001) 28–32
49. O'Grady, M.J., O'Hare, G.M.P.: Just-in-time multimedia distribution in a mobile computing environment. IEEE Multimedia 11(4) (2004) 62–74
50. O'Hare, G.M.P., O'Grady, M.J.: Gulliver's genie: A multi-agent system for ubiquitous and intelligent content delivery. Computer Communications 26(11) (2003) 1177–1187
51. Hagras, H., Callaghan, V., Colley, M., Clarke, G., Pounds-Cornish, A., Duman, H.: Creating an ambient-intelligence environment using embedded agents. Intelligent Systems 19(6) (2004) 12–20
52. Association for Computing Machinery ACM – http://www.acm.org
53. Institute of Electrical & Electronic Engineers (IEEE) – http://www.ieee.org
54. Keegan, S. O'Hare, G.M.P., O'Grady, M.J.: EasiShop: Ambient intelligence assists everyday shopping. Information Sciences 178(3) (2008) 588–611
55. Keegan, S., O'Hare, G.M.P.: EasiShop: Enabling uCommerce through intelligent mobile agent technologies. In: Horlait, E., magedanz, T., Glitho, R. (eds): Mobile Agents for Telecommunication Applications. Lecture Notes in Computer Science Vol. 2881. Springer (2003) 200–209

7

Automated Processing and Classification of Face Images for Human–Computer Interaction Applications

Ioanna-Ourania Stathopoulou and George A. Tsihrintzis

Department of Informatics, University of Piraeus, Piraeus 185 34, Greece,
iostath@unipi.gr, geoatsi@unipi.gr

Summary. Automated face detection and facial expression classification of images arises in the design of human–computer interaction and multimedia interactive service systems as a difficult, yet crucial, pattern recognition problem. Towards this goal, we have been building NEU-FACES, a novel system for processing multiple camera images of computer user faces to determine their affective state. In this chapter, we present an empirical study that we conducted to specify related design requirements, study statistically the expression recognition performance of humans, and identify quantitative facial features of high expression discrimination and classification power.

7.1 Introduction

7.1.1 The Importance of Understanding Emotions

According to psychologists, the fulfillment of emotional needs is essential and necessary to human well-being, as living with unmet emotional needs may cause pain, anxiety, depression, or violence eruptions [1–3]. Indeed, several of the best known problems that plagued human society in the twentieth century, such as drug and alcohol abuse or violence and criminality, derive from this inability to meet such basic emotional needs.

The first step in order for someone to fulfill his/her emotional needs is to be aware of them and recognize them and, next, to be able to meet them. Emotional needs are often categorized into two main categories: The first category consists of *emotional skill needs* and refers to awareness of emotions, both one's own and those of others, and the ability to manage them [4]. The second category is referred to as "*experiential emotional needs*" and tends to follow the Webster Dictionary definition of a need: "*A physiological or psychological requirement for the well-being of an organism.*" When one or more of these needs go unmet, an individual may suffer pain and, in extreme cases, chronic failure to meet these needs can have very severe effects.

I.-O. Stathopoulou and G.A. Tsihrintzis: *Automated Processing and Classification of Face Images for Human–Computer Interaction Applications*, Studies in Computational Intelligence (SCI) **104**, 107–136 (2008)
www.springerlink.com

Below are indicative lists of the two aforementioned categories of emotional skill and experiential needs. Specifically, emotional skill needs [3,4] are needs for basic skills and abilities for handling emotions, such as:

- *Emotional self-awareness*: a need to learn to appraise and express what one is feeling
- *Managing emotions*: the need to handle and regulate feelings so that they are appropriate
- *Self-motivation*: a need to learn to harness one's emotions in the service of a goal, for example by delaying gratification
- *Affect perception*: a need to accurately appraise what others are feeling as they are feeling and expressing it
- *Empathy*: a need to learn to appreciate what others are feeling (closely linked in the literature to emotional self-awareness)
- *Handling relationships*: primarily via managing the emotions of others. This skill is a necessary component of friendship, intimacy, popularity, and leadership

Experiential emotional needs [5,7] are mostly inherently social needs and are, therefore, usually met only with the assistance or presence of others. These include needs:

- For attention, which is strong and constant in children and fades to varying degrees in adulthood
- To feel that one's current emotional state is understood by others, particularly during strong emotional response
- To love and feel reciprocity of love
- To express affection and feel reciprocated affection expressed
- For reciprocity of sharing personal disclosure information
- To feel connected to others
- To belong to a larger group
- For intimacy
- To feel that one's emotional responses are acceptable by others
- To feel accepted by others
- To feel that emotional experience and responses are "normal"
- For touch, to be touched
- For security

7.1.2 How Advanced Human Computer Interaction Techniques Can Help

Technological advances may help people meet their emotional needs, at least during a human–computer interaction session. Although computers cannot replace interpersonal relations, they can assist humans to fulfill their needs. Such a case may arise, for example, during e-learning, where the teacher is not present, but the encouragement of the student or the reward is needed.

Indeed, computers offer great potential for supporting human emotional needs, because of the abilities of modern computational media. Specifically, interactive media:

- Are increasingly portable, smaller, cheaper, therefore they are increasingly able to be with their users at all times
- Soon they will be able to sense emotion via a variety of traditional means such as facial expression, tone of voice, and gesture
- Are able to be eternally attentive, particularly valuable for applications with young children
- Humans sometimes tend to treat them as real people

Interactive media can not only support educational needs and enable social interaction, but can also help people to partially meet their emotional needs. Some of such opportunities are identified below.

Supporting Emotional Skill Needs

Meeting someone's emotional needs is very important during a human–computer interaction process, especially in educational technology. This way, the computer program can enable learners to acquire academic skills and knowledge. It is conceivable that similar tools can be designed to address emotional skill needs. Software tutors could be built today for students of any age to learn about emotions; other tools could help build emotional awareness and management skills.

Emotional self-awareness is one of the basic emotional skill needs and a system able to recognize and record a person's needs is basic to modern human computer interaction techniques. A simple way to build such a system is by prompting the user to record emotions, possibly via selecting from a list of pre-defined emotions at random moments of the day. Work on such a tool is in investigation by the authors at the MIT Media Laboratory [1].

The other important part is real-time emotion sensing and recognition. This can be done, either by facial expression recognition, speech recognition, and gesture recognition or by combining two or three of these techniques. The realization of this technology may represent a fundamental advancement in human–computer interaction. For example, it may enable the development of an emotion-sensitive "Active Listener." Active listening is a simple but powerful skill used extensively by experienced therapists, and involves providing non-judgmental feedback, often about a speaker's emotional expression during conversation. While such a tool would probably rely on still-primitive speech processing capabilities, the potential benefit for such a tool is enormous.

Supporting Experiential Needs

While it is commonly assumed that experiential needs can only be met by other humans, this perception is not entirely true. People may satisfy several of these needs via other means, such as pet dogs or cats. In fact, people are able to establish relationships with a wide variety of organisms of various degrees of interactivity. During a human–computer interaction session, the system should be able to provide a bonding with the user and enable him to emotionally express himself. This, as an example, can be seen in recent products that feature computational simulations of pets demonstrate that interactive media can stimulate pet-like emotional bonding for children and adults alike [1]. Again, this conceptualization does not suggest that machines would substitute for interpersonal or even inter-organism contact, but offers a dramatic expansion in the availability and interactivity of non-human companions.

Speech and gesture recognition and facial expression classification can help the machines to meet human emotional needs. It is clear that humans can and do meet many of their experiential emotional needs on a daily basis using speech or facial expressions. Computational media has much to offer today and in the future to assist in their provision, but such products require research and development.

7.1.3 The Role of Facial Expression Recognition

Facial expressions play a significant communicative role in human-to-human interaction and interpersonal relations, because they can reveal information about the affective state, cognitive activity, personality, intention and psychological state of a person. It is common experience that the variety in facial expressions of humans is large and, furthermore, the mapping from psychological state to facial expression varies significantly from human to human. These two facts can make the analysis of the facial expressions of another person difficult and often ambiguous.

When mimicking human-to-human communication, human–computer interaction systems must determine the psychological state of a person, so that the computer can react accordingly. Indeed, images that contain faces are instrumental in the development of more effective and friendly methods in multimedia interactive services and human computer interaction systems. Vision-based human–computer interactive systems assume that information about a user's identity, state and intent can be extracted from images, and that computers can then react accordingly. Similar information can also be used in security control systems or in criminology to uncover possible criminals. Indeed, the facial expressions "neutral," "smile," "sad," "surprise," "angry," "disgust" and "bored-sleepy" arise very commonly during a typical human–computer interaction session and, thus, vision-based human–computer interaction systems that recognize them could guide the computer to "react" accordingly and attempt to better satisfy its user needs.

The task of processing facial images generally consists of two steps: (1) a face detection step in which the system determines whether or not there are any faces in an image and, if so, return the location and extent of each face and (2) a facial expression classification step, in which the system attempts to recognize the expression formed on a detected face. These problems are quite challenging because faces are non-rigid and have a high degree of variability in size, shape, color and texture. Furthermore, variations in pose, facial expression, image orientation and conditions add to the level of difficulty of the problem. The task is complicated further by the problem of pretence, i.e., the case of someone's facial expression not corresponding to his/her true psychological state.

Towards achieving the automated facial image processing goal, we have been developing a novel automated facial expression classification system [8–15], called NEU-FACES, in which features extracted as variations between the neutral and other common expressions are fed into neural network-based classifiers. Specifically, NEU-FACES is a two-module system, which automates both the face detection and the facial expression process. In the following sections of this chapter, we conduct an detailed study of automated processing and classification of face images for human–computer interaction applications. Specifically, we address the problems of face detection (Sect. 7.2), present an empirical study of the process of facial expression classification process by humans (Sect. 7.3), select features for use in automated facial expression classification (Sect. 7.4), design novel eye detection algorithms (Sect. 7.5), obtain quantitative estimates of the discrimination power of selected features (Sect. 7.6), present feature extraction procedures (Sect. 7.7), and build artificial neural network-based classifiers (Sect. 7.8). Furthermore, face detection and facial expression classification results are presented in Sect. 7.9, while conclusions are drawn and future work directions are identified in Sects. 7.10 and 7.11, respectively. Finally, key bibliographic references to literature relevant to this chapter are given at the end of this chapter.

7.2 Face Detection Module

7.2.1 General Information

The goal of face detection is to determine whether or not there are any faces in a given image and, if so, return the location and extent of each face. This problem is quite challenging because faces are non-rigid and have a high degree of variability in size, shape, color and texture. Furthermore, variations in pose, facial expression, image orientation and conditions add to the level of difficulty of the problem.

To address this problem, a number of works have appeared in the literature [e.g., 16–24] in which three main approaches can be identified. These approaches use (1) correlation templates, (2) deformable templates and (3) image invariants, respectively. In the first approach, we compute a difference measurement between one or more fixed target patterns and the candidate image locations and the output is thresholded for matches. The deformable templates of the second approach are similar in principle to correlation templates, but not rigid. To detect faces, in this approach we try to find mathematical and geometrical patterns that depict particular regions of the face and fit the template to different parts of the images and threshold the output for matches. Finally, in approaches based on image invariants, the aim is to find structural features that exist even when pose, viewpoint and lighting conditions vary, and then use them to detect faces.

Most of the aforementioned methods limit themselves to dealing with human faces in front view and suffer from several drawbacks: (1) they cannot detect a face that occupies an area smaller than 50×50 pixels or more than three faces in complex backgrounds or faces in images with defocus and noise problems or faces in side view and (2) they cannot address the problem of partial occlusion of mouth or wearing sunglasses. Although there are some research results that can address two or three of the aforementioned problems, there is still no system that can solve all of them.

7.2.2 Proposed Face Detection Algorithm

Our proposed algorithm falls within the third approach mentioned above. Specifically, we defined certain image invariants and used them to detect faces by feeding them into an artificial neural network. These image invariants were found based on the 14×16 pixel ratio template, proposed by P. Sinha [25,26], which we present below.

P. Sinha's Template

The method proposed by P. Sinha [25] (also included in [26]) combines template matching and image invariant approaches. P. Sinha aimed at finding a model that would satisfactorily represent some basic relationships between the regions of a human face. To be more specific, he found out that, while variations in illumination change the individual brightness of different parts of faces (such as eyes, cheeks, nose and forehead), the relative brightness of these parts remains unchanged. This relative brightness between facial parts is captured by an appropriate set of pair-wise brighter–darker relationships between sub-regions of the face.

We can see the proposed template in Fig. 7.1, where we observe 23 pair-wise relationships represented by arrows. The darker and brighter parts of the face are represented by darker and brighter shades of grey, respectively.

Our proposed face detection algorithm is built on this model. We preprocess a candidate image in order to enhance the relationships mentioned above and then, use the image as input to an artificial neural network. Then, the neural network will determine whether or not there is a face in the image.

The Algorithm

The main goal is to find the regions of the candidate image that contain human faces. The proposed system uses Artificial Neural Networks and operates in

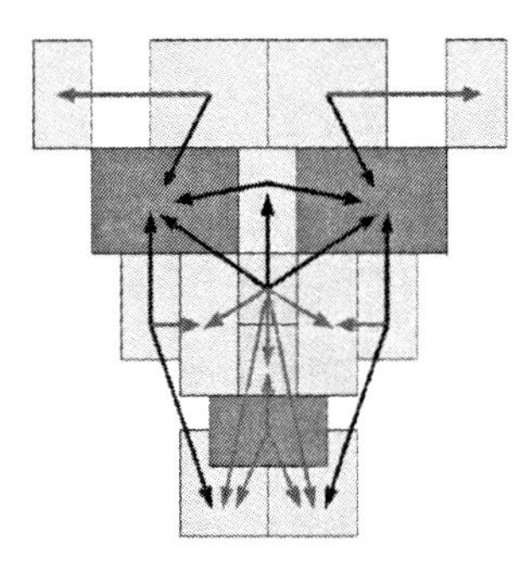

Fig. 7.1. The P. Sinha's template

two modes: training the neural network and using it to detect faces in a candidate image.

To train the neural network, we used a set of 285 images of faces and non-faces. We tried to find images of non-faces that are similar to human faces, so some of the non-face images contained dog, monkey and other animal "faces." These images where gathered from sources of the World Wide Web [27] and preprocessed before entered into the neural network.

To detect faces in a candidate image we apply a window, which scans the entire image, and preprocess each image region, the same way we preprocessed the images of the training set. Specifically, our algorithm works as follows:

1. We load the candidate image. It can be any three-dimensional (color) image
2. We scan through the entire image with a 35×35 pixel window. The image region defined by the window constitutes the "window pattern" for our system, which will be tested to determine whether it contains a face. We increase the size of the window gradually, so as to cover all the possible sizes of a face in a candidate image.
3. We preprocess the "window pattern"
   3.1 We apply Histogram Equalization techniques to enhance the contrast within the "window pattern."
   3.2 We compute the eigenvectors of the image using the Principal Component Analysis and the Nystrom Algorithm [28–31] to compute the normalized cuts.
   3.3 We compute three clusters of the image using the k-means algorithm and color each cluster with the average color.
   3.4 We convert the image from colored to grayscale (two-dimensional).
4. We resize the processed image into a dimension of 20×20 pixels and use it as input to the artificial neural network, which we present in the next section.

We summarize the basic stages of image processing in Table 7.1.

Table 7.1. The stages of the preprocess of the widow pattern

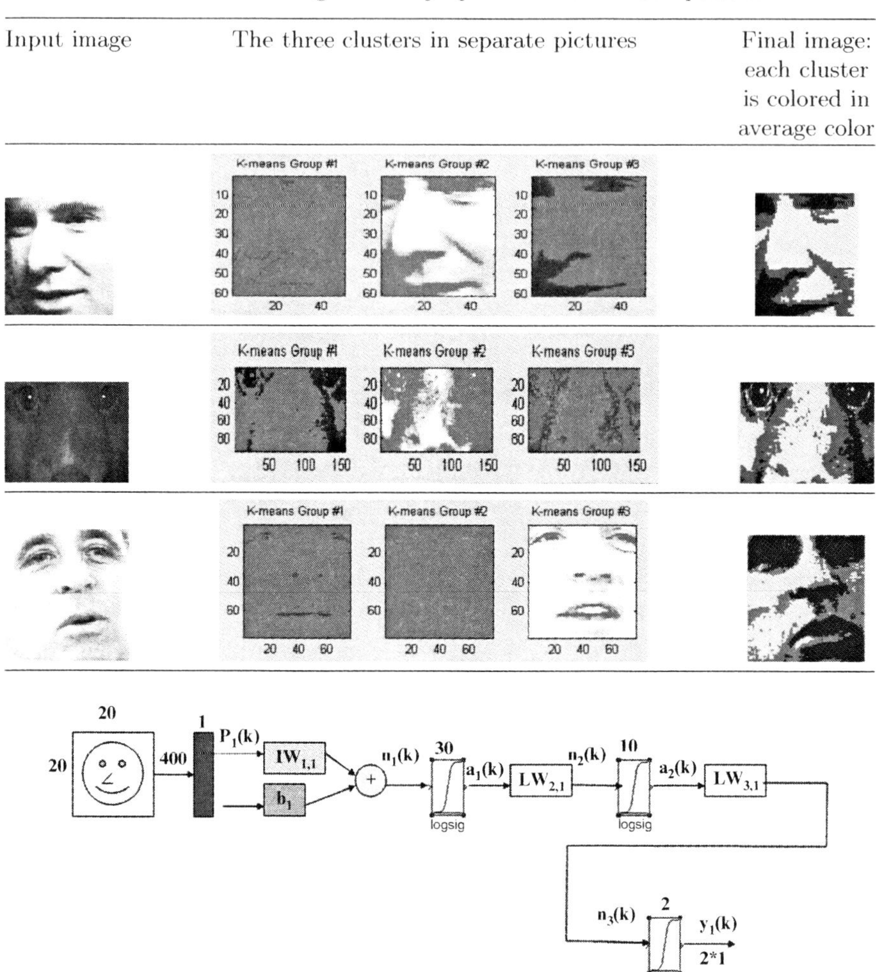

Fig. 7.2. The first ANN's structure

7.2.3 The Artificial Neural Network Structures

To classify window patterns as "faces" or "non-faces," we developed two different artificial neural networks, which are presented next.

The First Artificial Neural Network

This network takes as input the entire window pattern and produces a two-dimensional vector output. As seen in Fig. 7.2, the network consists of three hidden layers of thirty, ten and two neurons respectively, while its input

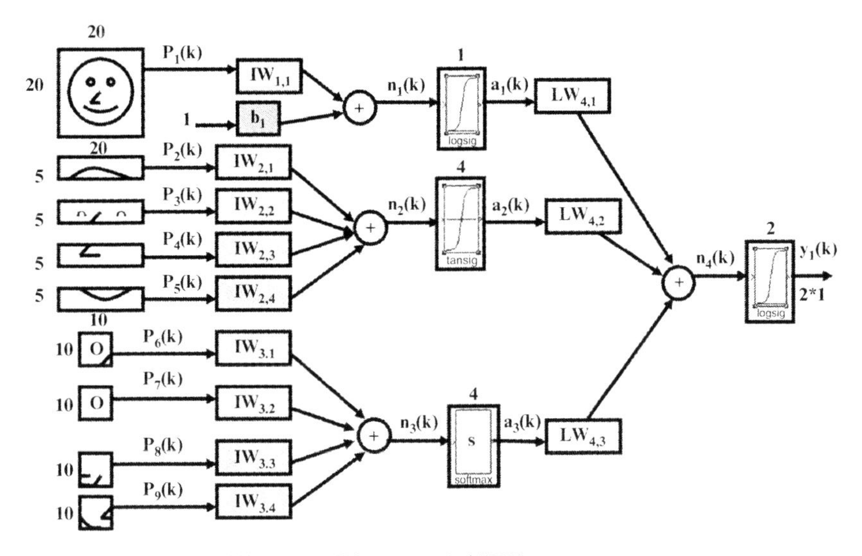

Fig. 7.3. The second ANN's structure

(window pattern) is of dimension 20-by-20 pixels. The neural network classifies the window pattern as "face" or "non-face." The output vector is of dimension 2-by-1 and equals to [1;0] if the window pattern represents a face or [0;1], else wise.

The Second Artificial Neural Network

The second neural network has four hidden layers with one, four, four and two neurons, respectively, and is fed with the following input data: (1) the entire "window pattern" (20-by-20 pixels), (2) four parts of the "window pattern," each 10-by-10 pixels and (3) another four parts of the "window pattern," 5-by-20 pixels (Fig. 7.3).

Each of the three types of inputs is fed into different hidden layers of the network. The first, second, and third sets of inputs are fed into the first, second, and third hidden layer, respectively, while the output vector is the same as for the first network. Clearly, the first network consists of fewer hidden layers with more neurons and requires less input data compared to the second.

7.2.4 Performance Evaluation

To train these two networks, we used a common training set of 285 images of faces and non-faces. During the training process, the first and second networks reached an error rate of 10^{-1} and 10^{-10}, respectively.

We required that, for both networks, the output vector value be close to [1;0] when the window pattern represented a face and [0;1] otherwise. This means that the output vector corresponds to the degree of membership of the image in one of the two clusters: "face-image" and "non-face-image." Some results of the two neural networks can be seen in Table 7.2.

The first network, even though it consisted of more neurons than the second one, did not detect faces in the images to a satisfactory degree, as did the second network. On the other hand, the execution speeds of two networks are comparable. Therefore, the second network was found superior in detecting faces in images. Some results of the face detection system are depicted in Fig. 7.4.

Table 7.2. Results of the two neural networks for various images

Input image	Pre-processed window pattern	First ANN's output	Second ANN's output
Face images			
		[0.5; 0.5]	[0.947; 0.063]
		[0.6; 0.4]	[1; 0]
		[0.5; 0.5]	[0.9717; 0.0283]
Non-face images			
		[0.5; 0.5]	[0; 1]
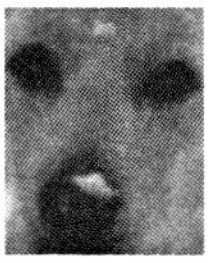		[0.5; 0.5]	[0; 1]
		[0.5; 0.5]	[0; 1]

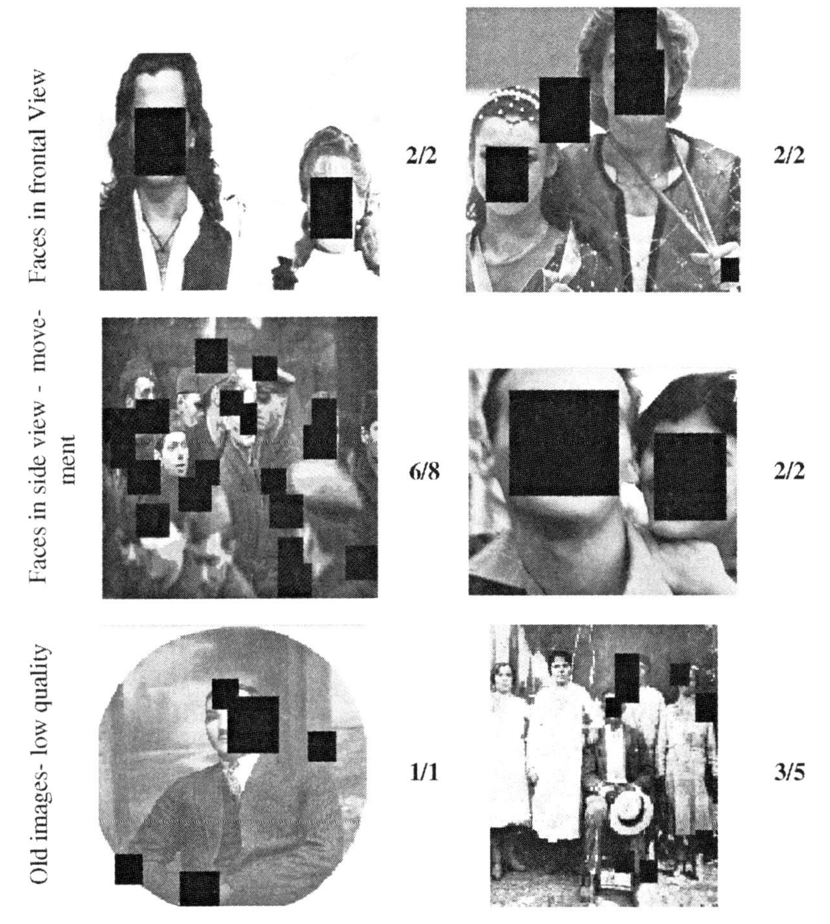

Fig. 7.4. Application of our face detection algorithm on three groups of images. The first group consists of images of faces in front view, while in the second group we have images blurred by motion, faces in side view and/or partially occluded, and complex backgrounds. Finally, in the third group we have old images of low quality

7.3 Empirical Study of Facial Expression Classification by Humans

7.3.1 General Information

After face detection, we must overcome the problem of facial expression recognition. In order to set some rules and understand how human expressions are been morphed and what they reveal about someone's psychological state, an empirical study was made. The scientific literature shows that, to address this problem, researchers have developed either *face feature*-based methods or methods which rely on *image-based representations of the face.* The former approaches rely on basic, image-extracted facial features and their location or computed relations between them, while the latter approaches utilize the entire face as input to artificial neural network-based classifiers.

For use in the development, training, and testing of facial expression classifiers, appropriate extensive facial databases are required. These databases are non-trivial to create, as they need to be sufficiently rich in both facial expression variety and representative samples of each expression. Moreover, the creators of the database need to make sure that the human models form their true facial expressions when posing. In the past years, only a relatively small number of relevant face databases have been presented in the literature. These include: (1) The AR Face Database [32], which contains over 4,000 color images of 126 persons' faces in front view, forming different facial expressions under various illumination conditions and occlusion (e.g., sun glasses and scarf). The main disadvantage of this database is its limitation to containing only four facial expressions, namely "neutral," "smile," "anger," and "scream." (2) The Japanese Female Facial Expression (JAFFE) Database [33] contains 213 images of the neutral and six additional basic facial expressions, as formed by ten Japanese female models. (3) The Yale Face Database [34], which contains 165 gray-scale GIF-formatted images of 15 individuals. These correspond to 11 images per subject of different facial expression or configuration, namely, center-light, with glasses, happy, left-light, without glasses, normal, right-light, sad, sleepy, surprised, and wink. (4) The Cohn-Kanade AU-Coded Facial Expression Database [35], which includes approximately 2,000 image sequences from over 200 subjects and is based on the Facial Action Coding System (FACS), first proposed by Paul Ekman [36]. (5) The MMI Facial Expression Database [37], which includes more than 1,500 samples of both static images and image sequences of faces in front and side view, displaying various expressions of emotion and single and multiple facial muscle activation.

Although many of the aforementioned face databases were considered for the development of our system (NEU-FACES), either the number of different facial expressions or the number of representative samples of them were found insufficient for developing NEU-FACES up to a fully operational form and, thus, we decided to create our own facial expression database. In this chapter, we present an empirical study of the facial expression classification problem in images, as well as details of the process followed in creating our facial expression database.

7.3.2 Our Study of Expression Classification by Humans Consisted of Three Steps

1. *Observation of the user's reactions during a typical human–computer interaction session*: From this step, we concluded that the facial expressions corresponding to the "neutral," "smile," "sad," "surprise," "angry," "disgust" and "bored-sleepy" psychological states arose very commonly in

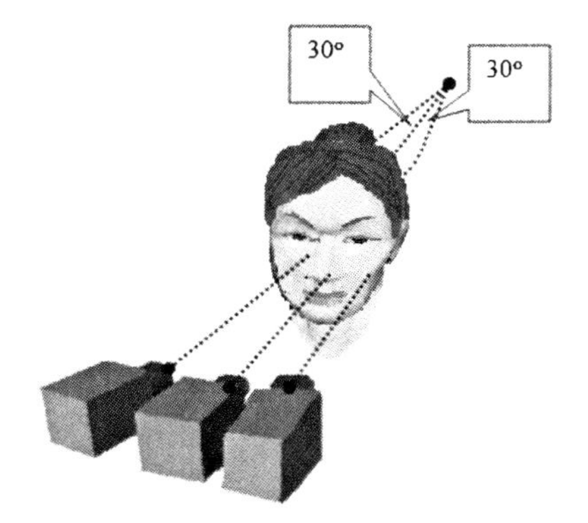

Fig. 7.5. The geometry of the data acquisition setup

human–computer interaction sessions and, thus, form the corresponding classes for our classification task

2. *Data acquisition*: To acquire image data, we built a three-camera system, as in Fig. 7.5. Specifically, three identical cameras of 800×600 pixel resolution were placed with their optical axes on the same horizontal plane and successively separated by 30-degree angles. Subjects were asked to form facial expressions, which were photographed by the three cameras simultaneously.
   To ensure spontaneity, the subject was presented with pictures on a screen behind the central camera. These pictures were expected to generate such emotional states that mapped on the subject's face as the desired facial expression. For example, to have a subject assume a "smile" expression, we showed him/her a picture of funny content. We photographed the resulting facial expression and only then asked him/her to classify this expression. If the image shown to him/her had resulted in the desired facial expression, the corresponding photographs were saved and labeled; otherwise, the procedure was repeated with other pictures. The final dataset consisted of 250 different persons, each forming the seven expressions: "neutral," "smile," "sad," "surprise," "angry," "disgust" and "bored-sleepy."
3. *Questionnaires – Classification of expressions by humans*: To understand how humans classify facial expressions and estimate the corresponding error rate, we developed a questionnaire in which each we asked 300 participants to classify the facial expressions in 36 images. Each participant could choose from 11 of the most common facial expressions, such as: "angry," "smile," "neutral," "surprise," etc., or specify some other expression he/she thought appropriate.

Specifically, our dataset consisted of three subsets of images, typical examples of which are shown in Table 7.3:

- Various images of individuals placed in a background and mimicking an expression

Table 7.3. Typical face image subsets in our questionnaire

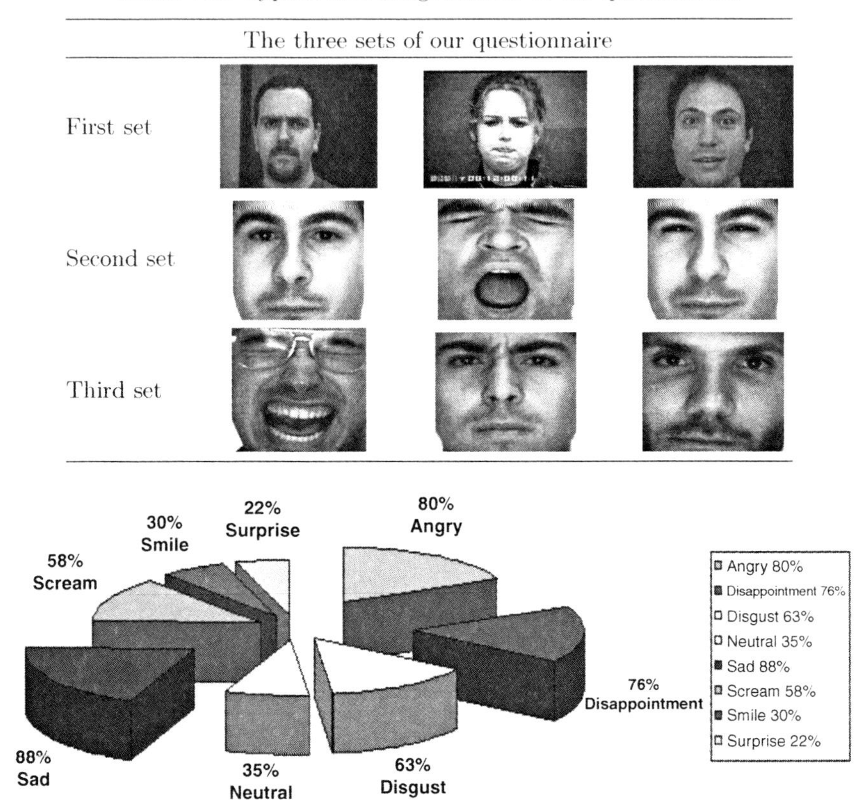

Fig. 7.6. Error rates in recognizing the expressions in our questionnaire

- A sequence of facial expressions of the same person without a background
- Facial images of various individuals without a background

We found that the "surprise" expression was the one recognized with the lowest error rate of 22%. Next, follow the "smile" and "neutral" expressions, with error rates of 30 and 35%, respectively. The highest error rate corresponded to the "sad" expression and rose to 88%, while the "angry" and "disappointment" expressions had an error rate of 80 and 76%, respectively. These are summarized in Fig. 7.6. From these findings, we conclude that the facial image classification task is quite challenging and the fact, that expressions such as "angry," "sad" and "disappointed" seem to differ significantly from person to person or some people may be too shy to form them clearly, may result in high classification error rates.

7.4 Feature Selection

From the collected dataset, we tried to identify differences between the "neutral" expression of a model and its deformation into other expressions, as typically highlighted in Table 7.4. To convert pixel data into a higher-level representation of shape, motion, color, texture and spatial configuration of the face and its components, we locate and extract the corner points of specific regions of the face, such as the eyes, the mouth and the brows, and compute their variations in size, orientation or texture between the neutral and some other expression. This constitutes the *feature extraction process* and reduces the dimensionality of the input space significantly, while retaining essential information of high discrimination power and stability.

Table 7.4. Differences among facial expressions

Variations between facial expressions			
Smile		Bored-sleepy	
	• Bigger-broader mouth • Slightly narrower eyes • Changes in the texture of the cheeks • Occasionally, changes in the orientation of brows		• Head slightly turned downwards • Eyes slightly closed • Occasionally, wrinkles formed in the forehead and different direction of the brows
Surprised		Sad	
	• Longer head • Bigger-wider eyes • Open mouth • Wrinkles in the forehead (changes in the texture) • Changes in the orientation of eyebrows (the eyebrows are raised)		• Changes in the direction of the mouth • Wrinkles formed on the chin (different texture) • Occasionally, wrinkles formed in the forehead and different direction of the brows
Angry		Disgusted-disapproving	
	• Wrinkles between the eyebrows (different textures) • Smaller eyes • Wrinkles in the chin • The mouth is tight • Occasionally, wrinkles over the eyebrows, in the forehead		• The distance between the nostrils and the eyes is shortened • Wrinkles between the eyebrows and on the nose • Wrinkles formed on the chin and the cheeks

7.5 Eye Detection

7.5.1 General Information

In vision-based intelligent multimodal human–computer interaction systems, eye feature detection is the key component in a facial expression recognition module, because eye features are the most salient features of a face. Eye features include pupil centers and their radii, eye corners and eyelid contours. Among these features, pupil centers and radii are the simplest to detect and estimate. So, the first detected face features are the eyes. Based on their location, we then detect the remaining features.

The development of a fully automated eye detection and feature extraction system, capable of fast and accurate eye feature estimation (e.g., [38–52]), is quite challenging. Some of the challenges that have to be addressed in developing such a system arise from the facts that faces are non-rigid and have a high degree of variability in size, shape, color and texture. Furthermore, variations in pose, image orientation and conditions add to the level of difficulty of the problem.

The precise eye corner location is crucial in three-dimensional face recognition (Blanz and Vetter 2003) and also determines the approximate eyelid contour location. Deformable contour models are well-known methods to extract object contour information in computer vision applications. In [40,41] made an attempt to utilize such algorithm in eye contour extraction. Because of its limitations under the conditions in [42], the deformable contour model is not optimal and stable in eye contour extraction. Specifically, the main difficulty lies with the fact that deformable models need careful formulation of the energy term and close model initialization in order to avoid unwanted contour results.

To overcome the limitations of deformable models, researchers have paid more attention to several landmark points of the eyes rather than extracting the entire continuous eye contour. The set of landmark points in the eye contour can then be fitted by mathematical functions [44–46]. These approaches generally result in better robustness, but reduced accuracy.

In this section, we present a novel eye detection and feature extraction algorithm, which we developed for use in our facial expression classification system and has been found quite accurate on a wide variety of images, regardless of illumination, pose, facial expression or even partial face occlusion caused by wearing glasses. The eye detection is applied on the detected face image, which is the result of our face detection module presented above. Some results can be seen in the Table 7.5.

7.5.2 The Eye Detection Algorithm

After successful face detection, we apply our eye detection algorithm in the upper 60% of the detected face region. We do not process the lower 40% of the face region to decrease the complexity of the algorithm and the required

Table 7.5. Face detection

Face detection	
Original image	Detected face

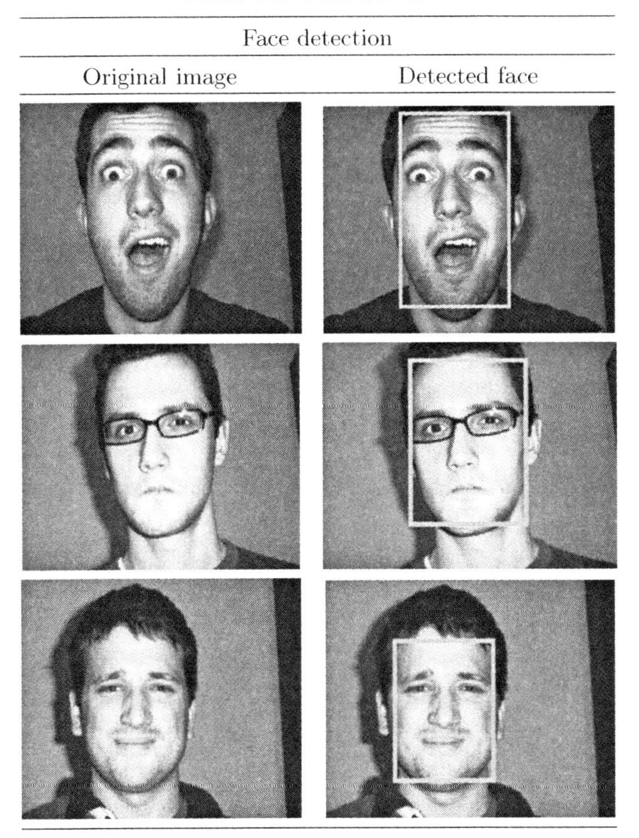

computational effort. The 60% of the upper detected face is selected based on our studies so as to cover cases of face rotation.

For better accuracy, the algorithm computes two different binary images, called "skin map" and "clustered image," respectively, and uses them to detect the eyes. Specifically, the algorithm follows four main steps:

- Skin extraction
- K-means clustering
- Combination of the resulting images and morphological processing
- Feature extraction

Skin Extraction

The skin filter is based on the Fleck and Forsyth algorithm [53]. The input color image must be in RGB format with color intensity values ranging from 0 to 255. The algorithm, works as follows:

1. The RGB image is transformed to log-opponent values I, Rg, and By, given by the Fleck and Forsyth algorithm, as follows:
   - $I = L(G)$
   - $Rg = L(R) - L(G)$
   - $By = L(B) - (L(G) + L(R))/2$

   The L(x) operation is defined as $L(x) = 105 * \log_{10}(x + 1)$. The log transformation makes the Rg and By values, as well as differences between I values (e.g., texture amplitude), independent of illumination level.
2. After filtering the Rg and By matrices, a texture amplitude map is used to find regions of low texture information. Usually, the skin is very smooth, so the skin regions are those with little texture.
3. In these selected areas, we further select the skin region based on the measures of hue and saturation, so as their color matches that of skin. The acceptable values of hue and saturation, are $110 \leq \text{hue} \leq 180$ and $0 \leq \text{saturation} \leq 130$, respectively.
4. A binary skin map is drawn, where if the pixel in the original image is in the same coordinates as the pixel map is skin, it is represented with 1, or 0 otherwise. The skin map array can be considered as a black and white binary image with skin regions appearing as white.

The resulting "skin map" usually represents the eyes, brows, nostrils, hair and other objects on the face (e.g., glasses), with white regions and the skin with black. Some results after applying the algorithm are shown in Table 7.6.

K-Means Clustering

On the same 60% of the detected face region, we compute three clusters of the image, color each cluster with the corresponding average color, and, finally, convert it to binary. The resulting image is called "clustered face image" and, as the "skin map," represents the eyes, brows, nostrils, hair and other objects on the face (e.g., glasses), as white regions and the skin as black. Some results, corresponding to the faces in the skin extraction step, are shown in Table 7.6.

Morphological Operations-Combining Two Images

On the two resulting images, we apply morphological operations. Our aim is to remove all other objects and to end up with a binary image of only the eyes, so as to make the eye detection task simpler. The algorithm, works as follows:

1. We apply a window, which clears the boundary pixels in the input image (usually representing the hair and the nostrils)
2. We remove areas whose size is too small. This removes some very small objects on the face (e.g., scars)
3. We remove the areas whose length is larger than the 1/3 of the total row size. In this step, wide areas, e.g., the skeleton of the glasses are removed.

Table 7.6. Eye extraction

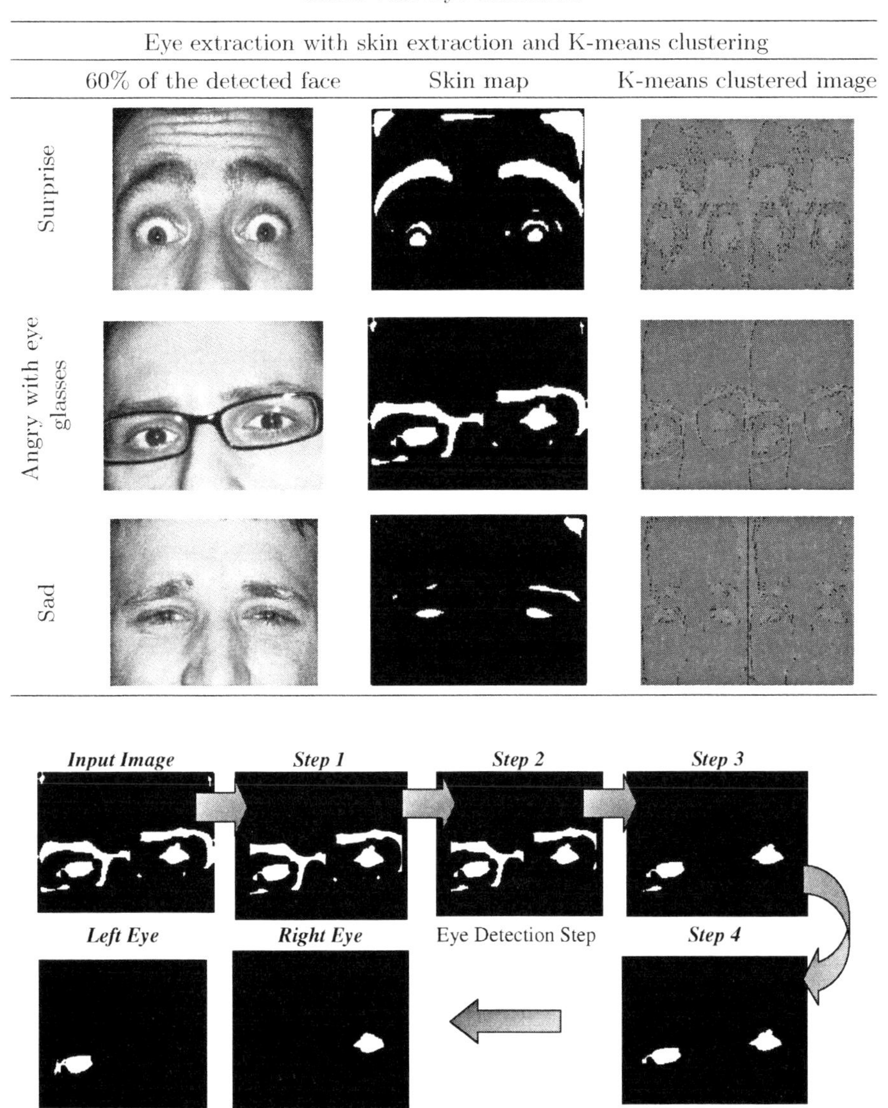

Fig. 7.7. Morphological operations

4. Finally, we remove the areas whose length is very small to remove other small objects of the face.

The image resulting at each step of the aforementioned algorithm is shown in Fig. 7.7, in which the input image was the skin map of the second image (i.e., the image of the where the person wearing glasses). The result is an image containing only the two eyes.

Table 7.7. Extracted features

Surprised 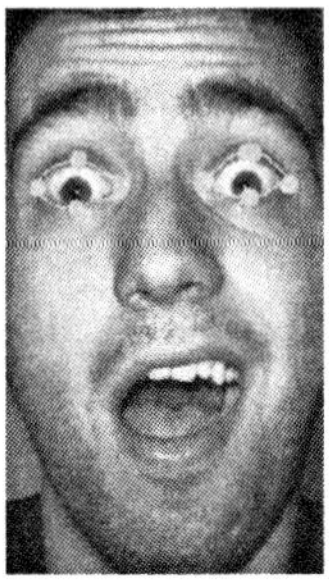	Angry wearing eye glasses 	Sad 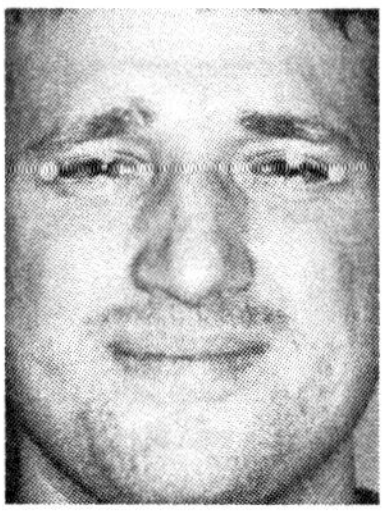

The detection is done on the two images. The final location of the eyes is found based on characteristics of the detected areas in the two images, e.g., the relative position of the eyes and their size, in relation to the original image. Each eye is depicted to a binary image, where the features are extracted next.

Feature Extraction

After detecting the eyes, we end up with two binary images representing the left and the right eye, respectively. First, we trace the outline of the eye area. Non-zero-valued pixels are assigned to an object and zero-valued pixels constitute the background. The curve is drawn based on these relative values.

To compute edge points, first we find the minimum and the maximum coordinates of the computed contour. Then, the center points are computed. Finally, the edge points are computed based on their relationships relatively to the center points and the contour. Finally, the coordinates of the extracted features are drawn in the original (color) image of the original size. The extracted edge points, in the original image, are depicted in Table 7.7.

7.5.3 Results

Our algorithm managed to locate and extract the edge points and curves of the eyes quite satisfactorily. It was tested on 900 images of 150 persons, each forming six facial expressions. It was found to be quite accurate on a wide variety of images, regardless of illumination, pose, facial expression, or even partial facial occlusion caused by wearing glasses. Of course, occasional errors appeared in more extreme cases, such as the case of eyes partially occluded by hair or when they are completely shut.

Our algorithm was further applied to a facial expression analysis system with good results. Some further examples from applying our algorithm on various images are shown in Fig. 7.8. In this picture, we demonstrate cases of

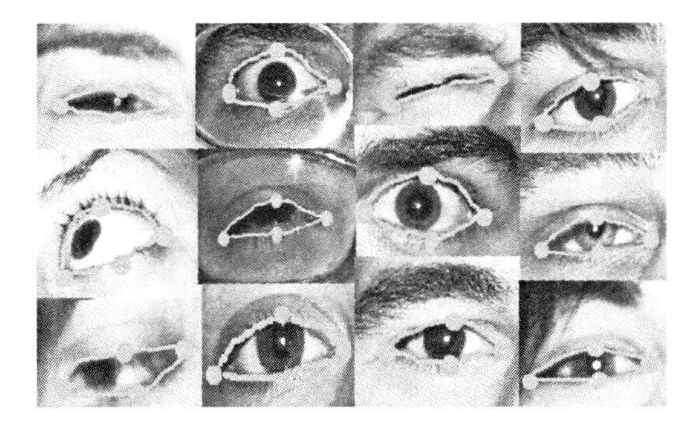

Fig. 7.8. Eye detection-feature extraction results

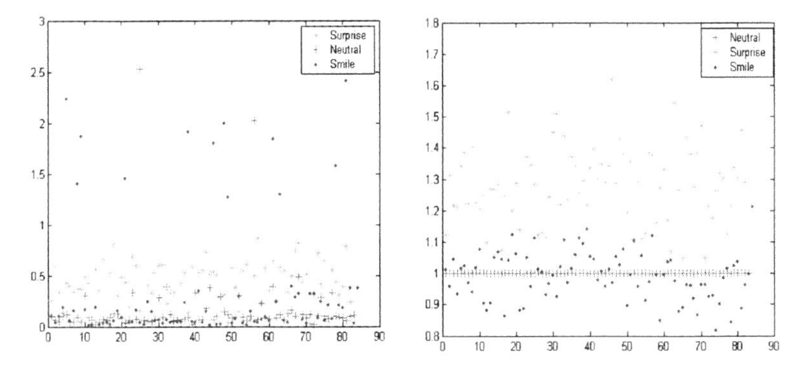

Fig. 7.9. Distribution of mouth (*left*) and face (*right*) dimension ratio

different illumination, facial expression, pose and partial facial occlusion by glasses.

7.6 Quantification of Feature Discrimination Power

We have found that a number of high discrimination power features may correspond to the location and shape of face components, as these vary significantly between the "neutral," "smile" and "surprise" facial expressions. For example, the distribution of the mouth and face dimension ratios over a broad range of face images are shown in Fig. 7.9 for these three different facial expressions. Similarly, the orientation and texture of specific portions of the face also vary significantly between expressions.

As far as the collected *side view* images are concerned, our study showed that formation of some expressions involves deformation of a person's head sides and, thus, additional classification features may be derived from side view face images. In fact, features may be more evident in side view rather than front view images for certain expressions. For example, better discrimination between the "neutral" and "smile" expression seems to be achieved in front view images, whereas the "surprise" expression seems to be better identified in side view images. Similarly, the "sad" and angry" are better discriminated in front rather than in side view images as forehead texture, one of the corresponding classification features, is better computed in front rather than side view images. Thus, we conclude that better facial expression classification results can be achieved by using images of several views of a person's face.

Furthermore, the texture and the orientation of specific portions of the face vary significantly between the expressions. Typical results of these two measures are shown in Tables 7.8 and 7.9.

Table 7.8. Different measures of the facial expressions

Measures of texture				
Region between the brows				
Expressions	Input image	Difference between the relevant "neutral expression"	Texture measure	Possible facial expression class
Neutral	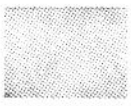		0	"neutral"
Smile			16	"neutral," "smile," "surprise"
Surprise			6	"neutral," "smile," "surprise"
Angry			44	"angry," "disgust"
Disgust	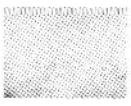	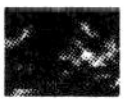	175	"angry," "disgust"
Forehead				
Neutral			0	"neutral," "smile," "angry," "disgust"
Smile			0	"neutral," "smile," "angry," "disgust"
Surprise			8	"surprise," "angry"
Angry			0	"neutral," "smile," "angry," "disgust"
Disgust			0	"neutral," "smile," "angry," "disgust"

Table 7.9. Different measures of the facial expressions

Measures of orientation				
Mouth orientation			Brow orientation	
First person	Neutral	1.8159°	Neutral	25.4796°
	Sad	45°	Sad	−61.2104°
Second person	Neutral	2.001°	Surprise	0.77162°
	Sad	25.4612°		
Third person	Neutral	0.3168°		
	Sad	5.9384°		
Measures of texture				
Forehead				
	Input Image	Processed (binary) image	Texture measure	
Neutral			0	
Surprise			23	
Region between the brows				
Neutral			1	
Angry			33	

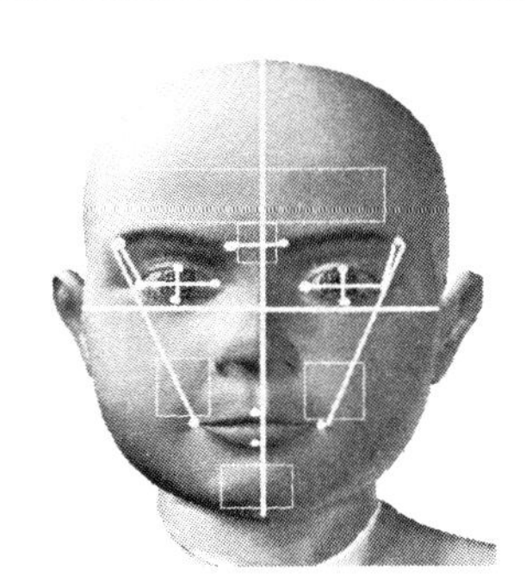

Fig. 7.10. The extracted features (*gray points*), the measured dimensions (*gray lines*) and the regions (*orthogonals*) of the face

7.7 Feature Extraction

The feature extraction process in NEU-FACES converts pixel data into a higher-level representation of shape, motion, color, texture and spatial configuration of the face or its components. We extract such classification features based on observations of facial changes that arise during formation of various facial expressions, as indicated in Fig. 7.10. Specifically, we locate and extract the corner points of specific regions of the face, such as the eyes, the mouth and the brows, and compute variations in size or orientation from one expression to another. Also, we extract specific regions of the face, such us the forehead or the region between the eyebrows, so as to compute their variations in texture.

Specifically, the feature extraction algorithm works as follows:

1. Search the binary face image and extract its parts (eyes, mouth and brows) into a new image of the same dimensions and coordinates as the original image.
2. In each image of a face part, locate corner points using relationships between neighboring pixel values. This results in the determination of 18 facial points which are subsequently used to form the classification feature vector.
3. Based on these corner points, extract the specific regions of the faces (e.g., forehead, region between the eyebrows). The extracted corner points and regions can be seen in the third column in Table 7.10, as they correspond to the five facial expressions of the same person shown in the first column. Although these regions are located in the binary face image, their texture measurement is computed from the corresponding region of the detected face image ("window pattern") in the second column.

Table 7.10. Face detection and feature extraction

	Input image	Detected face	Extracted features	Expression classification
Neutral				[*1.00*; 0.00; 0.00; 0.00; 0.00]
Smile				[0.16; *0.83*; 0.00; 0.01; 0.00]
Surprise			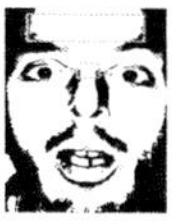	[0.01; 0.02; *0.92*; 0.05; 0.00]
Angry				[0.00; 0.00; 0.11; *0.63*; 0.36]
Disgust	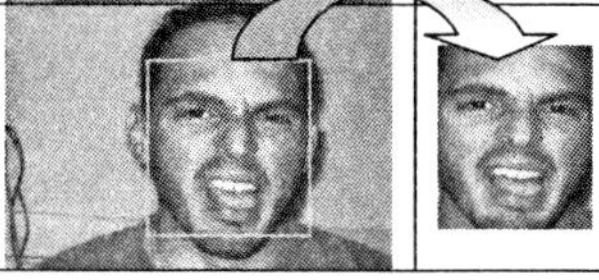			[0.00; 0.00; 0.08; 0.28; *0.64*]

4. Compute the Euclidean distances between these points, depicted with gray lines in Fig. 7.10, and certain specific ratios of these distances. Compute the orientation of the brows and the mouth. Finally, compute a measure of the texture for each of the specific regions based on the texture of the corresponding "neutral" expression.
5. The results of the previous steps form the feature vector which is fed into a neural network.

7.8 Facial Expression Classifier

In order to classify facial expressions, we developed a two layer artificial neural network which is fed with the input data: (1) left eye dimension ratio, (2) right eye dimension ratio, (3) mouth dimension ratio, (4) face dimension ratio, (5) forehead texture, (6) texture between the brows, (7) left eye brow direction, (8) right eye brow direction, and (9) mouth direction The network produces a five-dimensional output vector which can be regarded as the degree of membership of the face image in each of the "neutral," "smile," "surprise," "angry," "disgust-disapproval" and "bored-sleepy" classes. An illustration of the network architecture can be seen in Fig. 7.11.

7.9 Results

After computing the feature vector, we use it as input to an artificial neural network to classify facial images according to the expression they contain. Some of the results obtained by our neural network can be seen in Table 7.10. Specifically, in the first column we see a typical input image, whereas in the second column we see the results of the Face Detection Subsystem. The extracted features are shown in the third column and finally the Facial Expression Classification Subsystem's response is shown in the fourth column.

According to the requirements set, when the window pattern represented a "neutral" facial expression, the neural network should produce an output value of [1.00; 0.00; 0.00; 0.00; 0.00] or so. Similarly, for the "smile" expression, the output must be [0.00; 1.00; 0.00; 0.00; 0.00] and so on for the other

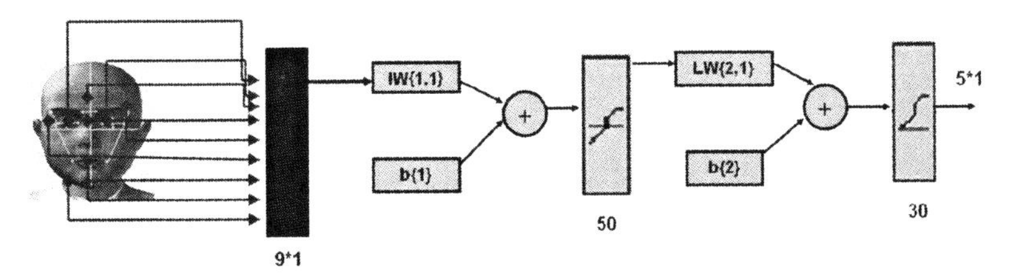

Fig. 7.11. The neural network-based facial expression classifier

expressions. The output value can be regarded as the degree of membership of the face image in each of the "neutral," "smile," "surprise," "angry," "disgust-disapproval" and "bored-sleepy" classes.

7.10 Conclusions

Automatic face detection and expression classification in images is a prerequisite for the development of novel human–computer interaction modalities. However, the development of integrated, fully operational such detection/classification systems is known to be non-trivial, a fact that was corroborated by our own statistical/empirical results regarding expression classification by humans. Towards building such systems, we developed a neural network-based system, called NEU-FACES, which first determines whether or not there are any faces in given images and, if so, returns the location and extent of each face. Next, we described features which allow the classification of several facial expressions and presented neural network-based classifiers which use them. The proposed system is fully functional and integrated, in that it consists of modules which capture face images, estimate the location and extent of faces, and classify facial expressions. Therefore, the present or improved versions of our system could be incorporated into advanced human–computer interaction systems and multimedia interactive services.

7.11 Future Work

In the future, we will extend this work in the following three directions: (1) we will improve our system by using wider training sets so as to cover a wider range of poses and cases of low quality of images, (2) we will investigate the need for classifying into more than the currently available facial expressions, so as to obtain more accurate estimates of a computer user's psychological state. In turn, this may require the extraction and tracing of additional facial points and corresponding features, (3) we will adapt our system so as to make it operational within the framework of mobile telephony, in which the quality of the input images is too low for existing systems to operate reliably.

Another extension of the present work of longer term interest will address several problems of ambiguity concerning the emotional meaning of facial expressions by processing contextual information that a multimodal human–computer interface may provide. For example, complementary research projects are being developed [54–56] that address the problem of emotion perception of users through their actions (mouse, keyboard, commands, system feedback) and through voice words. This and other related work will be presented on future occasions.

Acknowledgement

This work has been sponsored by the General Secretary of Research and Technology of the Greek Ministry of Development as part of the PENED-2003 basic research program.

References

1. J. Klein, R.W. Picard, and J. Riseberg. Support for Human Emotional Needs in Human–Computer Interaction, CHI'97 Workshop on Human Needs and Social Responsibility, 1997
2. M. Csikszentmihalyi. Flow: The Psychology of Optimal Experience, Harper and Row, New York, 1994
3. D. Goleman. Emotional Intelligence, Bantam Books, New York, 1995
4. P. Salovey and J.D. Mayer. Emotional Intelligence. Imagination, Cognition and Personality 9(3), 185–211 (1990)
5. A. Ortony, G.L. Clore, and A. Collins. The Cognitive Structure of Emotions, Cambridge University Press, 1988
6. H.W. Marsh and R.J. Shavelson. Self-concept: Its multifaceted, Hierarchical Structure. Educational Psychologist 20, 107–204 (1985)
7. M. Rosenberg. Conceiving the Self, Basic Books, New York, 1979
8. I.-O. Stathopoulou and G.A. Tsihrintzis. A new neural network-based method for face detection in images and applications in bioinformatics. Proceedings of the 6th International Workshop on Mathematical Methods in Scattering Theory and Biomedical Engineering, September 17–21, 2003
9. I.-O. Stathopoulou and G.A. Tsihrintzis. A neural network-based facial analysis system. 5th International Workshop on Image Analysis for Multimedia Interactive Services, Lisboa, Portugal, April 21–23, 2004
10. I.-O. Stathopoulou and G.A. Tsihrintzis. An improved neural network-based face detection and facial expression classification system. IEEE International Conference on Systems, Man, and Cybernetics 2004, The Hague, The Netherlands, October 10–13, 2004
11. I.-O. Stathopoulou and G.A. Tsihrintzis. Pre-processing and expression classification in low quality face images. 5th EURASIP Conference on Speech and Image Processing, Multimedia Communications and Services, Smolenice, Slovak Republic, June 29–July 2, 2005
12. I.-O. Stathopoulou and G.A. Tsihrintzis. Evaluation of the discrimination power of features extracted from 2-D and 3-D facial images for facial expression analysis. 13th European Signal Processing Conference, Antalya, Turkey, September 4–8, 2005
13. I.-O. Stathopoulou and G.A. Tsihrintzis. Detection and expression classification systems for face images (FADECS), 2005 IEEE Workshop on Signal Processing Systems (SiPS'05), Athens, Greece, November 2–4, 2005
14. I.-O. Stathopoulou and G.A. Tsihrintzis. An accurate method for eye detection and feature extraction in face color images, IWSSIP-2006. 13th International Conference on Systems, Signals and Image Processing, Budapest, Hungary, September 21–23, 2006

15. I.-O. Stathopoulou and G.A. Tsihrintzis. Facial expression classification: specifying requirements for an automated system. 10th International Conference on Knowledge-Based & Intelligent Information & Engineering Systems, Bournemouth, United Kingdom, October 9–11, 2006
16. A.J. Colmenarez and T.S. Huang. Face detection with information-based maximum discrimination. Computer Vision and Pattern Recognition 782–787 (1997)
17. G. Yang and T.S. Huang. Human face detection in a complex background. Pattern Recognition 27(1), 53–63 (1994)
18. S.Y. Lee, Y.K. Ham, and R.H. Park. Recognition of human front faces using knowledge-based feature extraction and neuro-fuzzy algorithm. Pattern Recognition 29(11), 1863–1876 (1996)
19. T.K. Leung, M.C. Burl, and P. Perona. Finding faces in cluttered scenes using random labeled graph matching. Fifth International Conference on Computer Vision, IEEE Computer Society Press, Cambridge, MA, pp. 637–644, 1995
20. H.A. Rowley, S. Baluja, and T. Kanade. Rotation invariant neural network-based face detection. CMU-CS-97-201, 1997
21. H.A. Rowley, S. Baluja, and T. Kanade. Neural Network-based face detection. IEEE Transactions on Pattern Analysis and Machine Intelligence, 20(1) (1998)
22. P. Juell and R. Marsh. A hierarchical neural network for human face detection. Pattern Recognition 29(5), 781–787 (1996)
23. C. Lin and K. Fan. Triangle-based approach to the detection of human face. Pattern Recognition 34, 1271–1284 (2001)
24. K.K. Sung and T. Poggio. Example-based learning for view-based human face detection. Proceedings on Image Understanding Workshop, Monterey, CA, pp. 843–850, 1994
25. P. Sinha. Qualitative representations for recognition. Lecture Notes in Computer Science, Springer, Berlin Heidelberg New York, LNCS 2525, pp. 249–262, 2002
26. M.–H. Yang and N. Ahuja. Face Detection and Gesture Recognition for Human–Computer Interaction. Kluwer Academic Publishers, Dordrecht, The Netherlands, 2001
27. Gender Classification (Databases). http://ise0.stanford.edu/class/ee368a_proj00/project15/intro.html; http://ise0.stanford.edu/class/ee368a_proj00/project15/append_a.html
28. C. Fowlkes, S. Belongie, and J. Malik. Efficient Spatiotemporal Grouping Using the Nystr om Method CVPR, Hawaii, December 2001
29. S. Belongie. Notes on Clustering Pointsets with Normalized Cuts, 2000
30. J. Shi and J. Malik. Normalized cuts and image segmentation. IEEE Transactions on Pattern Analysis and Machine Intelligence 22(8) (2000)
31. Image Segmentation using the Nystrom Method. http://rick.ucsd.edu/~bleong/
32. A.M. Martinez and R. Benavente. The AR face database. CVC Tech. Report #24, 1998. http://cobweb.ecn.purdue.edu/~aleix/aleix_face_DB.html
33. M.J. Lyons, J. Budynek, and S. Akamatsu. Automatic classification of single facial images. IEEE Transactions on Pattern Analysis and Machine Intelligence 21(12), 1357–1362 (1999) http://www.kasrl.org/jaffe.html
34. The Yale Database: http://cvc.yale.edu/projects/yalefaces/yalefaces.html
35. T. Kanade, J.F. Cohn, and Y. Tian. Comprehensive database for facial expression analysis. Proceedings of the 4th IEEE International Conference on Automatic Face and Gesture Recognition, Grenoble, France, 2000. http://vasc.ri.cmu.edu/idb/html/face/facial_expression/

36. P. Ekman and W. Friesen. Unmasking the face: A Guide to Recognizing Emotions from Facial Expressions, Prentice Hall, Englewood Cliffs, NJ, 1975. http://www.paulekman.com/
37. M. Pantic, M.F. Valstar, R. Rademaker, and L. Maat. Web-based Database for Facial Expression Analysis. Proc. IEEE Int'l Conf. Multimedia and Expo (ICME'05), Amsterdam, The Netherlands, July 2005. http://www.mmifacedb.com/
38. J. Ahlberg. A system for face localization and facial feature extraction. Tech. Rep. LITH-ISY-R-2172, Linkoping University, 1999
39. J. Huang and H. Wechsler. Eye detection using optimal wavelet packets and RBFs. International Journal of Pattern Recognition Artificial Intelligence (IJPRAI) 13(6), 1009–1026 (1999)
40. K.M. Lam and H. Yan. Locating and extracting the covered eye in human face image. Pattern Recognition 29(5), 771–779 (1996)
41. L. Yin and A. Basu. Realistic animation using extended adaptive mesh for model based coding. In: Proc. Energy Minimization Methods in Computer Vision and Pattern Recognition, pp. 315–318, 1999
42. V. Vezhnevets and A. Degtiareva. Robust and accurate eye contour extraction. In: Proc. Graphicon, pp. 81–84, 2003
43. M. Ladws, J.C. Vorbruggen, J. Buhmann, and J. Lange. Distortion invariant object recognition in the dynamic link architecture. IEEE Transactions Computational 42, 570–582 (1993)
44. T.S. Lee. Image representation using 2D Gabor wavelets. IEEE Transactions PAMI 18(10), 959–971 (1996)
45. G.C. Feng and P.C. Yuen. Variance projection function and its application to eye detection for human face recognition. Patter Recognition Letters 19, 899–906 (1998)
46. C. Feng and P.C. Yuen. Multi-cues eye detection on gray intensity image. Pattern Recognition 34(5), 1033–1046 (2001)
47. T. Goto, W.-S. Lee, and N. Magnenat-Thalmann. Facial feature extraction for quick 3D face modeling. Signal Processing: Image Communication 17, 243–259 (2002)
48. M. Porat and Y.Y. Zeevi. The generalized Gabor scheme of image representation in biological and machine vision. IEEE Transactions PAMI 10(4), 452–468 (1988)
49. K. Ville, J.-K. Kamarainen, and H. Kalviainen. Simple Gabor feature space for invariant object recognition. Patter Recognition Letters 25, 311–318 (2004)
50. A. Yulle, P. Hallinan, and D. Cohen. Feature extraction from faces using deformable templates. International Journal of Computation Vision 8(2), 99–111 (1992)
51. N. Zheng, W. Song, and W. Li. Image coding based on flexible contour model. Machine Graphics Vision 1(8), 83–94 (1999)
52. Z.-H. Zhou and X. Geng. Projection functions for eye detection. Pattern Recognition 37(5), 1049–1056 (2004)
53. M.M. Fleck, D.A. Forsyth, and C. Bregler. Finding Naked People, 1996 European Conference on Computer Vision, Vol. II, pp. 592–602, 1996
54. M. Virvou and E. Alepis. Mobile educational features in authoring tools for personalised tutoring. Journal Computers and Education 44(1), 53–68 (2005)

55. M. Virvou and G. Katsionis. Relating error diagnosis and performance characteristics for affect perception and empathy in an educational software application. Proceedings of the 10th International Conference on Human Computer Interaction (HCII) 2003, June 22–27 2003, Crete, Greece
56. M. Virvou and E. Alepis. Creating tutoring characters through a Web-based authoring tool for educational software. Proceedings of the IEEE International Conference on Systems Man & Cybernetics, 2003, Washington D.C., USA

8

On Developing and Communicating User Models for Distance Learning Based on Assignment and Exam Data

Thanasis Hadzilacos[1,2], Dimitris Kalles[1], and Christos Pierrakeas[1]

[1] Hellenic Open University, Patras, Greece, {thh, kalles, pierrakeas}@eap.gr
[2] Research Academic Computer Technology Institute, Patras, Greece

Summary. Students who enrol in the undergraduate program on informatics at the Hellenic Open University (HOU) demonstrate significant difficulties in advancing beyond the introductory courses. We use decision trees and genetic algorithms to analyze their academic performance throughout an academic year. Based on the accuracy of the generated rules, we analyze the educational impact of specific tutoring practices. We examine the applicability of these techniques in a senior course and reflect on some software engineering issues involved in the development of organization-wide measurement systems. All results are based on data drawn from academic records.

8.1 Introduction

All measurements affect that which is being measured. However, measurement for performance evaluation greatly affects what is being measured and may distort the process being evaluated. Such distortions make the education field particularly vulnerable due to misleading policies. For example the number of computers per 100 high school students is an indicator of educational ICT utilization (other things being equal). When however this indicator is used as a funding goal ("by 2007 all EC member states should have 1 computer per 10 students") it results in schools keeping old machines and in the total ICT budget being shifted toward hardware.

It is reasonable to suggest that student success is a natural success indicator of a University (of a teacher, of a class, or of a course). However, if that success is used as a criterion for tutor contract renewal, and if students must evaluate their own teachers, then tutors may tend to lax their standards. This chapter is about dealing with this issue in the context of a large (over 25,000 students) Open-and-Distance-Learning (ODL) University, the Hellenic Open University (HOU) and in particular its undergraduate Informatics program. We ask how we can detect best distance tutoring practices and associate them

T. Hadzilacos et al.: *On Developing and Communicating User Models for Distance Learning Based on Assignment and Exam Data*, Studies in Computational Intelligence (SCI) **104**, 137–155 (2008)
www.springerlink.com

with measures of students' success in an "objective" way and, subsequently, effectively disseminate this information to all interested parties.

The measurement strategy we have developed to-date in HOU has been progressively refined to deal with two closely linked problems: that of predicting student success in the final exams and that of analyzing whether some specific tutoring practices have any effect on the performance of students. Each problem gives rise to the emergence of a different type of user model. A student model allows us, in principle, to explain and maybe predict why some students fail in the exams while others succeed. A tutor model allows us to infer the extent to which a group of tutors diffuses its collective capacity effectively into the student population they advise. However, both types of models can be subsequently interpreted in terms of the effectiveness of the educational system that the university implements.

Appreciating the inherent caveats of measurement per se as well as those of measurement for performance evaluation and that any decision of how to interpret the models above inevitably raises ethical or political issues, we have first developed our models and the associated performance indicators on a small and controlled scale. At that selected scale, where we have a direct knowledge of most underlying problems faced by the tutors and the tactics employed by them as a response to these problems, we are able to form hypotheses and test them with reasonable creditworthiness. We have employed decision trees and genetic algorithms to develop such model hypotheses and indicators. All along we have designed our measurements with a view of the potential for their possible organizational adoption, since such adoption would need to scale into about 2,000 students groups (and their associated tutors) clustered within about 200 course modules. Keeping always in mind the scale and diversity of user groups is a key software engineering aspect of devising a system to disseminate performance findings.

This chapter is structured in five subsequent sections. First, we briefly review the problem of student performance at large in the context of distance learning and offer a very short review of our past approaches. We then describe the machine learning technologies that we have used in our experimentation. Following that, we present the results of our experiments in an incremental fashion, to reflect our increasing insight into the problem. We then discuss our findings and describe the implications of these findings, as well as reflect on their validity and their potential for wide-spread use. At the last section we conclude and touch on the impact aspects of our work.

8.2 Background on the Educational Domain

A module is the basic educational unit at HOU. It runs for about ten months and is the equivalent of about 3–4 conventional university semester courses. A student may register with up to three modules per year. For each module, a student is expected to attend five plenary class meetings throughout the academic year. A typical class contains about thirty students and is assigned to a tutor (tutors of classes of the same module collaborate on various course aspects). Class meetings are about four hours long and are structured along tutor presentations, group-work and review of homework. Furthermore, each student must turn in some written assignments (typically four or six), which contribute towards the final grade, before sitting a written exam.

Students fail a module and may not sit the written exam if they do not achieve a pass grade in the assignments they turn in; these students must repeat that module afresh. A student who only fails the written exam may sit it on the following academic year (without having to turn in assignments); such "virtual" students are also assigned to student groups but the tutor is only responsible for marking their exam papers.

The vast majority (up to 98%) of registered students in the "Informatics" program, upon being admitted at HOU, selects the module "Introduction to Informatics" (INF10). Following that, and according to university recommendations, they will typically select the modules "Fundamental Software Engineering" (INF11) and "Mathematics" (INF12). Those three can be considered as junior year modules and since they are the most heavily populated, they serve as test-beds for experimentation on drop-out issues. Collectively, these modules cover fundamental topics on mathematics, software engineering, programming, databases, operating systems and data structures.

Drop-out is a significant issue in ODL universities [1] and at HOU it mostly occurs very early in the studies, as a result of failure in a junior year module. Such failures skew the academic resources of the HOU system towards filtering the input rather than polishing the output, from a quantitative point of view. Even though this may be perfectly acceptable from an educational, political and administrative point of view (even from a social one) we must analyze and strive to understand the mechanism and the reasons of failure. This could significantly enhance the ability of HOU to fine-tune its tutoring and admission policies without compromising academic rigor.

Previous work at HOU on the subject of student dropout in the field of informatics showed that the total percentage of dropouts reaches 28.4%, compared to 71.6% who continue [2]. This is in line with similar findings reported elsewhere [3]. Generally, dropout is due to a number of reasons related to numerous factors [4]; such factors usually being the degree of difficulty of the selected courses and subject areas, the false estimation of time that could be dedicated to studies, health problems, etc. [5].

There are two key educational problems that have been identified as being core aspects of these failures. The first is that these courses are heavy on mathematics and adult students have not had many opportunities to sharpen their mathematical skills since high-school graduation (which has typically occurred at about 10 years prior to enrolling at HOU). The second is that the lack of a structured academic experience may have rendered dormant one's general learning skills and attitudes.

Analyzing the performance of such high-risk students is a goal towards achieving tutoring excellence. It is important to mention that the great majority of students dropped out after failing to deliver the first one or two written assignments. It is, thus, reasonable to assert that predicting a student's performance can enable a tutor to take early remedial measures by providing more focused coaching, especially in issues such as priority setting and time management. So, we have embarked on a data analysis effort to address this problem. Key demographic characteristics of students (such as age, sex, residence etc), their marks in written assignments and their presence or absence in plenary meetings may constitute the data for the task of explaining (and predicting) whether a student would eventually pass or fail a specific module.

Initial experimentation at HOU [6] consisted of using several machine learning techniques to predict student performance with reference to the final examination. The scope of the experimentation was to investigate the effectiveness and efficiency of machine learning techniques in such a context. The WEKA toolkit [7] was used because it supports a diverse collection of techniques (see Table 8.2 for which representative ones were eventually used). The key result was that learning algorithms could enable tutors to predict student performance with satisfying accuracy long before final examination. The key finding that lead to that result was that success in the initial written assignments is a strong indicator of success in the examination. Furthermore, our tutoring experience corroborates that finding.

We then employed the GATREE system [8] as the tool of choice for our experiments, to progressively set and test hypotheses of increasing complexity based on the data sets that were available from the university registry. The formation and development of these tests is the core content of this chapter and is presented and discussed in detail in the following sections. GATREE is a decision tree builder that employs genetic algorithms to evolve populations of decision trees; it was eventually used because it produces short comprehensible trees. Of course, GATREE was first used [9] to confirm the qualitative validity of the original findings experiments [6], also serving as result replication, before advancing to more elaborate experiments [10–12].

8.3 The Experimentation Toolkit

In our work we have relied on decision trees to produce performance models. Decision trees can be considered as rule representations that, besides being accurate, can produce comprehensible output, which can be also evaluated from a qualitative point of view [13]. In a decision tree nodes contain *test attributes* and leaves contain *class descriptors*.

A decision tree for the (student) failure analysis problem could look like the one in Fig. 8.1 and tells us that a mediocre grade at the second assignment (root) is an indicator of possible failure (left branch) at the exams, whereas a non-mediocre grade refers the alert to the fourth (last) assignment.

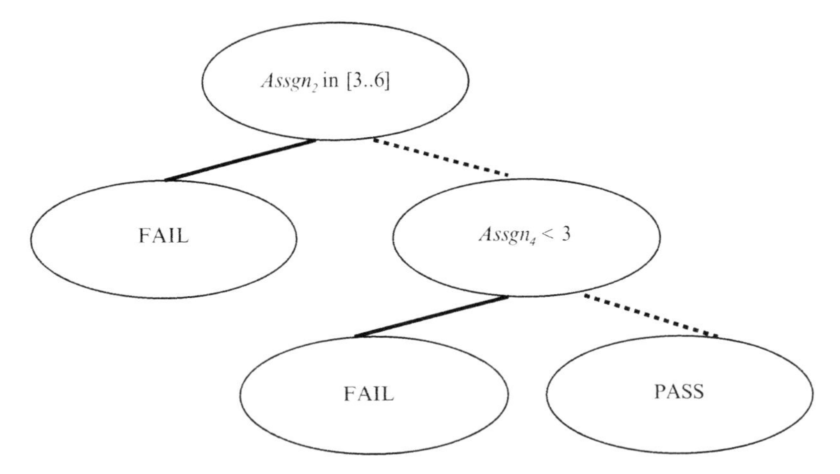

Fig. 8.1. A sample decision tree [9]

Table 8.1. A sample decision tree training set [9]

Assgn_1	Assgn_2	Assgn_3	Assgn_4	Exam
...	...	...	...	...
4.6	7.1	3.8	9.1	Pass
9.1	5.1	4.6	3.8	Fail
7.6	7.1	5.8	6.1	Pass
...	...	...	...	...

Decision trees are usually produced by analyzing the structure of examples (*training instances*), which are given in a tabular form. An excerpt of a training set that could have produced such a tree is shown in Table 8.1. Note that the three examples shown are consistent with the decision tree. As this may not always be the case, there rises the need to measure accuracy, even on the training set, in order to compare the quality of two decision trees which offer competing explanations for the same data set.

Note that the sample decision tree does not utilize data neither on the first nor the third assignments, but such data is shown in the associated table. Such *dimensionality reduction* information is typical of why decision trees are useful; if we consistently derive trees on some problem that seem to not use some data column, we feel quite safe to not collect measurements for that data column. Of course, simple correlation could also deliver such information, however it is the visual representation advantages of decision trees that have rendered them as very popular data analysis tools.

Genetic algorithms [13] can directly evolve binary decision trees. GATREE [8] does so by evolving populations of trees according to a fitness function that allows for fine-tuning decision tree size vs. accuracy on the training set (the example above is based on a tree similar to the ones produced by

GATREE for our experiments). At each time-point (in genetic algorithms dialect: *generation*) a certain number of decision trees (*population*) is generated and sorted according to some criterion (*fitness*). Based on that ordering, certain transformations (*genetic operators*) are performed on some members of the population to produce a new population. For example, a mutation may modify the test attribute at a node or the class label at a leaf, while a cross-over may exchange parts between decision trees.

The GATREE fitness function is $fitness(Tree_i) = CorrectClassified_i^2 * x / (size_i^2 + x)$. The first part of the product is the actual number of training instances that a decision tree (a member of a population) classifies correctly. The second part of the product (the size factor) includes a factor x which regulates the relative contribution of the tree size into the overall fitness; thus, the payoff is greater for smaller trees.

Our earlier work with WEKA [7] employed quite a few of the tools implemented therein. As far as decision trees are concerned, we used J48 (an implementation of the C4.5 algorithm [14]), J48-REP (a variant that implements reduced-error-pruning for decision trees [14]) and J48-Bagging (a variant that reduces variance [15]). We also used MLP (a standard neural network implementation [16]), 3-NN (a standard 3-Nearest-Neighbor), and also experimented with logistic regression [17] and naïve Bayes models.

When using GATREE, we used the default settings for the genetic algorithm operations and set cross-over probability at 0.99 and mutation probability at 0.01. When using WEKA we used the default settings provided by the application. However, *all* our experiments were carried out using tenfold cross-validation, on which all averages are based (i.e., one-tenth of the training set was reserved for testing purposes and the model was built by training on the remaining nine-tenths; furthermore, ten such stages were carried out by rotating the testing one-tenth).

8.4 The Experimental Results

We will now offer a structured presentation of the experimental results, as these were generated in the course of our experimentation with various modeling hypotheses.

8.4.1 Addressing the Fundamental Problem: Can Success (Failure) be Modeled?

The first experiments were conducted prior to our work [6] and are summarized here. The key observations were that learning algorithms could enable tutors to pinpoint students at risk of failure with satisfactory accuracy and that demographics seem to play a rather limited role (the variants of the experiments that were also based on demographic data did not show any significant deviation from those solely based on actual academic data).

These experiments were based on student data from the INF10 module and for the 2000–1 academic year and were rather unconventional in how they reported the accuracy of the induced models, since they did not follow the cross-validation approach described above. As far as techniques are concerned, they employed decision trees, neural networks, Naïve Bayes, instance-based learning (nearest neighbor), logistic regression and support vector machines [18].

In our new starting experiments we first used WEKA to bring these earlier results up-to-date, now based on cross-validation testing. We also reported the model size, for symbolic classifiers only, based on the full training data set; therein leaves were counted as nodes and trees were binary. The overall results are shown in Table 8.2.

The previous findings are indeed useful as they stand. Based on our tutoring experience, success in the initial written assignments is a strong indicator of successful follow-up. This is also corroborated by the fact that such success motivates the student towards a more whole-hearted approach to studying, based on the positive reinforcement. There is a risk here, of course, that the tutor might be tempted to artificially increase early grades; this could hide weaknesses by conveying the wrong message to the student.

We then employed GATREE and first tried a small-scale experimentation with up to 250 generations and up to 250 members per generation (see Table 8.3).

Where an X/Y number is reported as model size the interpretation is as follows: the genetic algorithm produced (at the end) a best tree of Y nodes, which was then post-processed by hand to eliminate redundant nodes (which

Table 8.2. Summary results for conventional classifiers [9]

Classifier	Average accuracy	Accuracy on training set	Model size
J48 (binary decision tree)	65.41	82.27	57
J48 (bdt – with REP)	69.48	75.00	33
J48 (bdt – with bagging)	69.48	86.92	N/A
MLP	66.86	93.02	N/A
Logistic regression	67.44	75.58	N/A
Naïve Bayes	63.08	65.12	N/A
3-NN	63.66	75.58	N/A

Table 8.3. Summary results for basic GATREE decision trees classifiers [9]

Classifier	Average accuracy	Accuracy on training set	Model size
(gen/pop) 250/250	77.06	87.50	*33*/37
(gen/pop) 250/100	80.00	86.63	39
(gen/pop) 100/250	78.53	86.05	*15*/17
(gen/pop) 100/100	76.76	84.59	19

Table 8.4. Summary results for extended GATREE decision trees classifiers [9]

Classifier	Average Accuracy	Accuracy on Training Set	Model Size
(gen/pop) 100/1000	75.59	86.92	41
(gen/pop) 250/1000	79.12	88.95	**43**/49
(gen/pop) 1000/1000	74.71	88.66	27
(gen/pop) 1000/250	79.12	88.95	25
(gen/pop) 1000/1000	80.00	92.15	51

Table 8.5. The experimental range for the success (failure) prediction problem

Year	INF10	INF11	INF12
2000–1	√	√	√
2001–2	√	√	√
2002–3	√	√	

sneak into the structure due to the absence of any post-processing) resulting in a tree of X nodes. To ensure the validity of the experimentation we used the same data sets of the original experimentation, including demographic data and quantized data. That experimental session confirmed that GATREE, when compared to conventional decision-tree classifiers, produces significantly more accurate trees. The tree size is comparable and the running time is longer, but due to its overall short duration, the exercise can be considered a success.

Following that, we scaled experiments up to 1,000 generations and up to 1,000 members per generation (see Table 8.4). The impressive finding was that for this type of classification problems the extended runs seemed to not be warranted.

These excellent findings paved the way for more extensive experimentation with more data sets and more modules (see Table 8.5).

The data sets detailed above were used for comparing the GATREE system and the J48 implementation available in WEKA. Each data set consisted of about 500 students records. We insisted on using decision trees because we believe that a small accurate model is a very important tool at the hands of a tutor, to assist in the task of continuously monitoring a student's performance with reference to the possibility of passing the final exam. A small model is easier to communicate among peers, easier to compare with competing ones and can have wider applicability. Quite importantly, the fact that we insisted on small models (by setting the x parameter of GATREE to its lowest allowable value) was justified by the results. Size-biased trees (with x at lowest value) almost always delivered a better performance compared to accuracy-biased trees (where the system first tries to score on accuracy and then progress towards smaller trees).

The comparison results demonstrated that GATREE models were very short and, often, more accurate than their conventionally induced counterparts (even if the latter ones employ pruning). There did exist a couple of data sets where the accuracy gap exceeded 2% (in favor of J48-REP). Overall, however, the consistent conciseness of the GATREE model coupled with its straightforward derivation and the ability to pay in time in order to "buy" more models (by simply increasing the number of generations and the populations) are key quality differentiators when considering the use of such a model at an operational scale, either to explain or to predict failure.

8.4.2 Forming and Testing Focused Hypotheses: Dealing with Drop-Out Students

Students who dropped out usually claimed [2] that they were not able to correctly estimate the time that they would have to devote to their professional activity and as a result, the time dedicated to their education decreased unexpectedly. One in four students also felt [2] that their knowledge was not sufficient for university-level studies (other reasons, such as family or health issues were also quoted). It is, thus, reasonable to assert that predicting a student's performance could enable a tutor to take early remedial measures by providing more focused coaching, especially in issues such as priority setting and time management.

As earlier described, it is the junior "virtual" students who are most prone to drop-out. Failure on a senior year course should simply postpone graduation as the fundamental commitment to studying has been already made. However, failure in a junior course, and for the HOU case and the Informatics program, this refers to the INF10, INF11 and INF12 modules, can contribute to a decision to drop out. This is because the learning investment is not yet large enough to warrant a certain attitude of persistence and because the student may not have had the time to familiarize oneself with the distance learning mode of education (which, given time, allows one to dovetail studying more effectively with other activities).

Virtual students are not entitled to attending plenary sessions, and to having their assignments graded by the group tutor (as a matter of fact they are not even requested to submit assignments). In practice this regulation may be relaxed by a tutor, who may opt to extend an invitation to these virtual students to attend some plenary sessions. Usually, all tutors of a module will either accept or decline to relax the regulation. Of course, there is no focused follow-up of the progress of virtual students, as opposed to the case with typical students.

Any attempt to address these realities involves a political decision that must necessarily take into account the university's administrative regulations.

One step taken by tutors of the INF10 and INF11 modules is to hold a plenary marking session of tutors for each module after an examination, and to discuss variations in individual marking styles based on a predefined assignment of points to exam questions. This is especially important for problems that involve design or prose argumentation. We note that this practice is not widespread within HOU.

Table 8.6. Summary description of approaches

Approach	INF10	INF11	INF12
Post-exam plenary marking session	√	√	
Grouping of virtual students		√	

A further ad hoc step taken (during the 2003–4 academic year) by the INF11 tutors was to group all virtual students in one group and assign one experienced tutor to that group, as opposed to the usual practice of distributing virtual students across tutors. These students were fully supported by an asynchronous discussion forum and by synchronous virtual classrooms. The tutor did neither hold a physical meeting nor correct any assignments. This was in line with the HOU regulations and, coincidentally, served as a convenient constraint on the "degrees of freedom" of the educational experiment.

A summary description of these policies is shown in Table 8.6.

We then tried to establish indicators on the effectiveness of these approaches.

Our methodology was to attempt to use the student data sets as input to GATREE to develop success/failure models represented as decision trees and then use the differences between the models derived when we omit some attributes to reflect on the importance of these attributes. This was similar to the approach employed in the original experimentation [6], where progressively narrower (with fewer attributes) data sets were used to argue about the eventual irrelevance of demographics.

Table 8.7 shows a sample of the extended data set for academic year 2001–2. It is similar to the one shown in Table 8.1, but contains two extra columns to account for which tutor group the student belongs to and at which year the student first enrolled. For example, the first complete data row tells us that a particular student was based in Athens (*Ath-*), and belonged to the third group at that city. The third complete data row tells us that the student from Herakleion (*Her-*) is a virtual student, since he took that particular course a year earlier than the majority (of course, there might be the case that the particular student may have postponed enrolment in that course, but the reader should by now appreciate the type of information that can be extracted from registry data).

We first tried to deal with the issue whether we might be able to obtain an overall (typical and virtual students included) model that deals with explaining (and, ultimately, predicting) exam success, across the three modules that have three distinct policies.

All data referred to the 2003–4 academic year. To appreciate the scale, we note that for modules INF10, INF11 and INF12 student populations are at about 1,000, 500 and 600 students respectively.

Table 8.8 summarizes the initial results on accuracy (measured in %), which are shown for experiments of 300 generations with 300 individuals per

Table 8.7. A sample decision tree extended training set

Assgn_1	Assgn_2	Assgn_3	Assgn_4	Year	Tutor group	Exam
...	...	...	...	...	...	...
4.6	7.1	3.8	9.1	2001	Ath-3	Pass
9.1	5.1	4.6	3.8	2001	Pat-1	Fail
7.6	7.1	5.8	6.1	2000	Her-2	Pass
...	...	...	...	...	...	...

Table 8.8. Accuracy comparison for GATREE decision trees (adapted from [10])

Data set	Accuracy – $x = 1{,}000$	Accuracy – $x = 10{,}000$
INF10: Basic	78.20	77.42*
INF10: Basic_T	78.20	77.30*
INF10: Basic_Y	83.60	84.61
INF10: Basic_TY	83.37	83.60
INF11: Basic	82.05	79.74*
INF11: Basic_T	81.28	80.26*
INF11: Basic_Y	82.31	84.36
INF11: Basic_TY	81.03	83.33
INF12: Basic	62.54	65.08
INF12: Basic_T	63.73	64.07
INF12: Basic_Y	70.51	72.03
INF12: Basic_TY	70.68	73.05

generation, for two values of x (basically experimenting with *allowable* small and large trees). We only report accuracies because (the eventually small) size was *not* affected. The *Basic* version of the training set consisted of all student records, where the only available attributes are the assignment grades and the class attribute is the *pass/fail* flag. The *Basic_T* version of the training set includes the tutor as an attribute, whereas the *Basic_Y* version includes as an attribute the year of first sitting the exam for that module. The *Basic_TY* version includes both additions (as shown in the sample of Table 8.7).

Starting from the INF11 module, we saw that the shorter trees ($x = 1{,}000$) were more consistent across the variants of the same data set. We argued that this had a twofold interpretation. On one hand, the larger trees were less well-fitted than the smaller ones (note the accuracy reduction for non-year-inclusive data-sets, as shown in starred numbers). This could well be an indication of over-fitting. On the other hand, it might suggest that the year attribute might have importance; this would concur well with the explanation that students who have failed to pass through the examination filter may be unlikely to have confidence to pursue their studies actively.

We then observed that, for INF12, the year attribute remained a top contributor to the model (all *_Y* versions did rather well). For INF10 and INF11 short trees again suggested that the year attribute was less important than for INF12, quite markedly so for INF11, where the year attribute was essentially suppressed. For larger trees, both for INF10 and INF11, the importance of the year attribute seemed to rise but at the expense of an overall reduction trend for the *Basic* models. This lent weight to the over-fitting argument but still did not rule out a potential importance of the year attribute, since one cannot easily wipe out the a priori disadvantage of virtual students.

However, the increase in accuracy for the INF10 models that used the year attribute was much less that the corresponding accuracy for the INF12 models. This observation combined with the observation that the average accuracies for INF10 were also larger than the average accuracies for INF12 were interpreted as an indicator that the plenary "marking" session of INF10 helped trim out potential grading inconsistencies. Of course, this might also be a contributor to the underlying quality of the INF11 models, but at the resolution level we were working, we could not easily confirm or refute the level of this contribution.

To do that we decided to cross-check the models using the tutor groups differences. The measurement was based on partitioned data sets. As opposed to treating each student population as an individual data set, we partitioned them into module groups, according to how tutors were assigned to groups. This partitioning allowed us to examine an important question: do the grading practices of one tutor apply predictably well to a group supervised by another tutor at the same module?

To answer the question, we devised a detailed study based on inducing models for one group and testing them at another. An example is shown in Table 8.9. D_i refers to the student group supervised by tutor i. CV_i refers to the cross-validation accuracy reported on training on data set D_i. $V_{i,j}$ refers to the validation accuracy reported on data set D_j, using the model of data set D_i.

We only focused on modules INF10 and INF11, where n takes the value of 31 and 16 respectively.

For each module, we first computed the above table at its basic version, using data formatted as shown in Table 8.1. We then computed the above table at its extended version, using additional data in the format shown in Table 8.7. We then subtracted the two tables (*basic* – *extended*) and computed the average of all cells. Table 8.10 shows the results, also including calculations for standard deviations and experiments with some accuracy/size trade-offs.

We noted that group models were clearly more aligned within the INF11 module which demonstrated "smoothing" across its training set partitions.

Table 8.9. A template for tabulating cross-testing results [11]

Data Set	D_1	D_2	...	D_n
D_1	CV_1	$V_{1,2}$	...	...
D_2	...	CV_2	...	...
...	...	...	...	...
D_n	...	...	...	...

Table 8.10. Accuracy differences [11]

Data set	Smallest trees	Allow larger trees if accuracy grows as well
INF10	−4.92 (10.39)	−3.65 (9.12)
INF11	0.28 (8.01)	0.24 (8.58)

This success indicator suggested that the failure explanation should be traced solely to academic performance (i.e., assignments) and that virtual students were not disadvantaged. The results shown in Table 8.8 were confirmed and, moreover, better reflected the differences between the INF10 and INF11 modules.

Summarizing the results, it seemed that when one focuses on limiting drop-out, the effective smoothing-out of the year-and-tutor factors in the success-failure model should benefit from a purely educational decision: by assigning an experienced tutor to directly deal with virtual students, besides enforcing a plenary marking session. Another alternative might be to train all tutors to be more active in discussion fora and more proficient in virtual classroom techniques, but this could demand a substantial mentality shift of the tutors and substantial vocational training resources.

8.4.3 Forming and Testing Focused Hypotheses: When Tutors do Matter

The previous experiments explored the question of how far one might be tempted by statistics to go in pin-pointing educational decisions that may impact the level of education offered to students. In the next round of experiments, we tried to investigate whether the above metrics could be used to pin-point differences in tutoring practices.

Again, we based our measurement in partitioned data sets, using the template in Table 8.9. For this round of experiments we used data formatted as shown in Table 8.1 (i.e., the *simple* format), and expanded our data with a senior year module (INF31), selected because one of the authors had a first hand experience in tutoring a group therein for the last four academic years (2006–7 included). For INF31, we had access to registry data from the 2004–5 academic year and we promptly used them. Note that, INF31 was a senior module and was less populated (six student groups accounting for a total of about 160 students). For this experiment, each table like Table 8.9 was summarized by the average value of its cells; the initial results are shown in Table 8.11.

Should we infer that senior course (INF31) tutors demonstrated a tighter homogeneity than their junior course (INF10, INF11) colleagues? Or, are the above findings the results of processes inherent (but, not yet identified) in the very different students populations (junior vs. senior)?

Table 8.11. Group cross-checking accuracy results [12]

Data set	INF31	INF10	INF11
Accuracy	92.05	83.60	82.31

Herein lurks the danger of using statistics (even, sophisticated) without adequate domain knowledge. To further analyze the above data we went a step further. We noted that the overall exam success rate in INF31 was nearly 100%, whereas in the other two modules (before factoring in the students who drop out) the success rate was substantially below 50%. That observation offered a new angle; few differences should be spotted between student groups that are overwhelmingly successful (as seems to be the case in INF31)!

The key to these results was in the fact that the term "exam" actually aggregated two sessions; a student sits the second exam if the first one is unsuccessful. In this light, we observed that the near 100% rate of INF31 was due to the second exam, whereas the success rate for the second exam in INF10 and INF11 was very small (compared to the overall rate). When we used the grades for the first INF31 exam, the 92.05 result dropped to 62.31 (use Table 8.11 as a benchmark).

The findings are telling (even without standard deviations). What initially appeared as homogeneity among tutors, subsequently evolved into a wide difference. While it might be tempting to reframe the previous question of homogeneity of tutoring practices, albeit in the opposite direction, we argued that the take home message was that of identifying the gap in cross-checking between groups of the same module as opposed to pitting modules against each other.

8.5 Discussion

The wider context of our research is to investigate the building of an "early warning and reaction system" for students with "weak" performance. As the cautious reader might imagine, user acceptance of such a system has delicate operational and political aspects that can easily transcend technical issues. To deal with this, we first need to think who the user might be.

Note that while we do not expect the individual or aggregate results to necessarily hold at other educational establishments, we believe that the measurement approach (with a relatively wide assortment of tools) should be readily applicable beyond our application case study. We also explore these directions below.

A sensitive point is that it would be unwise to simply consider the higher or lower overall *absolute* accuracy rate of (any) model in one module as an indicator of success of an approach, at least at this early stage of the research. It is for this reason that in the experiments described above we never pit one module's accuracy against another module's accuracy; besides referring to different student populations (including differences in population sizes), a module also refers to different tutors and to another scientific field.

An obvious plausible user is any student; therefore a key issue is how to communicate the model to the student. One option is to "publicize" the model in advance to the students. Even though the findings may not recur in the next year (though they might hold remarkably well [9]), an aura of "precaution" could be implicitly but effectively communicated, highlighting an otherwise obvious advice: unless one performs consistently well, a failure is quite likely. Another option suggests that the model should be only "publicized" to tutors, who then must make individual decisions how to use it. This can have some variances, however. For example, "using" the model can be both served by the tutor who ignores students who do not send in their first assignment and by the tutor who persistently goes with focused counseling after students who do badly in their third assignment. A middle-of-the-road course may be also possible.

The above discussion suggests that, besides the students, tutors are also candidate users. A very sensitive point is the interpretation of the individual model differences within each module. It might be tempting to think that this singles out groups within modules (and, as a result, their tutors) which seem to be not integrating into a module-wide view. This could have far-reaching effects if not properly managed. Unless, we painstakingly understand all background that may have led a tutor to adopt a particular approach (and, in that course, analyze all false alarms), pitting tutors' performance against each other is bound to create strife that will endanger the legitimacy of the approach. A full background analysis can be, however, safely ruled out due to the excessive "detective" costs that it entails. It is far more instructive to communicate the findings without any impact on performance evaluation and allow each individual tutor to reflect on their tutoring approach. It is very interesting that this is remarkably similar to the issue of whether we can actually use the models derived to better target each individual student.

Note that, in both cases, the issue of how one publicizes the measurement results can have profound differences in how these results are received and interpreted. We believe that any approach which might lead us into taking micro-managing decisions would create a distraction. What is more important, we claim, is to detect and observe the trends within the module itself and try to understand what macro-managing actions need to be taken at the module level.

We now explore whether this suggests that the university at large is a primary user.

We cannot yet answer whether the approach of the INF11 tutors is an approach that would have had replicable educational results in the other modules. The most obvious reason is that exact replication of the above experiments is impossible. Had we wanted to experiment with INF11 approach in INF10, we cannot hope to ever again observe the given set of students and their assignment to groups within modules, as well as the given set of tutors and their assignment to groups. This is one of the reasons that we progressively narrowed down our experiments: we started at only one undergraduate program, then focused on the most junior and well-subscribed modules, then singled out the two ones that demonstrated one difference only at the policy level and, finally, we slightly expanded our scope by including a well-understood senior module.

Note that the question of credibility of the results runs across all our experiments. In [9] we first set out an initial presentation of building classifiers for student performance using decision trees and genetic algorithms. In [10] we experimented with using the "virtual student" property as a discriminator of the classification – therein we demonstrated that when the discriminating property is suppressed, this is an indication of the fact that the second chance these students get is a substantial one. In [11, 12] we argued that the results were consistent over several experimental rounds with relatively small data sets; this is essential to boost the credibility of the derived models.

By establishing credible statistics at this level, we have finally stepped into working in the opposite direction, that of expanding our experimental range. This must be done across modules of the same program [12], then across programs, and at the same time, we must deal with the question of whether these models are consistent across consecutive academic years [9].

Our measurement approach is self-contained, in the sense that all data required are available at the university registry. This eliminates a substantial risk factor, that of having to collect the data from the tutors and then consolidate it. At the same time, however, we have not yet dealt with the software and data architectural aspects of collecting the data. While it is straightforward for an investigating team of several researchers to collect the data directly from the "source" for pilot projects, establishing an organization-wide process also entails issues of dealing with privacy issues (we must not disclose any individual student record).

Such issues will probably be affected by the decision of whether the models are computed centrally or in a decentralized fashion (by devolving responsibility to the tutors, for example). In any case, deploying our measurement scheme in an organization-wide context would also lend support to our initial preference for short models [9]. At the same time, the possibility of a decentralized scheme also suggests that we should strive to use tools that do not demand a steep learning curve on the part of the tutors. As a result we intend to continue favoring GATREE compared to other software for the particular data analysis tasks (of course, up-takers of our approach could decide using other tools). This will probably be a recurring theme in our attempts to diffuse our approach because conventional statistics [19, 20] can be cumbersome to disseminate to people with a background on humanities or arts, and this could have an adverse impact on the user acceptance of such systems.

The above all raise the fundamental question of whether one measures the performance of actors (students or tutors) or the performance of the system at large (the ODL system implemented in HOU). We strongly believe that the experiments documented in this chapter point towards the conjecture that is the latter alternative that has the most potential from an educational point of view.

We acknowledge that our approach may be at a wide tangent to policies as practiced by today's universities. We also acknowledge that (even) the strategy consensus required for fielding such systems may take indefinitely longer than the resolution of the technical and scientific issues that we have already identified. Actually, such consensus is probably of a political rather than a technological nature [21] and may account for the relative lack of related work [22].

Still, however, the emergence of tools and methodologies like the one we have proposed in this chapter, serves the need to equip educational stakeholders with the insight they direly need in order to contemplate policy alternatives and to reflect upon past decision based on actual data [23]. It may be that individual results struggle to reach the level of statistical significance, but it will always be up to innovative people to explore scenarios based on data interpretation to reflect about future policy directions.

8.6 Conclusions

We have presented how we have used an advanced AI method in order to obtain information necessary for the application of educational policies and we have explored the potential data flows from the data collection stage to the user presentation stage.

Quality control systems are fundamental in any large organization. Measurement is a *sine-qua-non* method of quality control and models are essential to explain or predict performance in quality control systems. We have argued that deciding what to measure and how to convey the measurement (and the underlying measurement message) in a university context is a very complex task, especially when the potential users can range from one (the university itself) to well over several hundreds (the tutors) and to tens of thousands (the students). To do this successfully, we always treat our experiments as an integral part of a software engineering project.

If we tread carefully and succeed, we expect that this will render our approach applicable to any educational setting where performance measurement can be cast in terms of test performance. Taking the sting out of individual performance evaluation but still being able to convey the full unabridged message should be a venerable target for every learning community.

Acknowledgements

This chapter shares setting-the-context paragraphs with related references [8–12], as the accurate description of the technical infrastructure and the educational environment is a significant factor in conveying our subsequent results precisely. This chapter is an invited extended version of a referenced conference paper [11], encompassing all previous development work and presenting how the authors' approach to-date has evolved based on the experiments and their results.

References

1. Open Leaning (2004). Special issue on "Student retention in open and distance learning". 19:1 (online).
2. Xenos, M., Pierrakeas, C., & Pintelas, P. (2002). A survey on student dropout rates and dropout causes concerning the students in the Course of Informatics of the Hellenic Open University. Computers & Education, 39, 361–377.
3. Cardon, P., & Christensen, K. (1998). Technology-based programs and dropout prevention. The Journal of Technology Studies, XXIV:1 (Winter/Spring – online).
4. Eisenberg, E., & Dowsett, T. (1990). Student drop – out from a distance education project course: A new method of analysis. Distance Education, 11:2.
5. Kaye, T., & Rumble, G. (1991). Open universities: A comparative approach. Prospects, 21:2, 214–216.
6. Kotsiantis, S., Pierrakeas, C., & Pintelas, P. (2004). Predicting students' performance in distance learning using Machine Learning techniques. Applied Artificial Intelligence, 18:5, 411–426.
7. Witten, I., & Frank, E. (2000). Data mining: practical machine learning tools and techniques with Java implementations. San Mateo, CA: Morgan Kaufmann.
8. Papagelis, A., & Kalles, D. (2001). Breeding decision trees using evolutionary techniques. Proceedings of the International Conference on Machine Learning, Williamstown, MA, pp. 393–400, Morgan Kaufmann.
9. Kalles, D., & Pierrakeas, Ch. (2006). Analyzing student performance in distance learning with genetic algorithms and decision trees. Applied Artificial Intelligence, 20(8), 655–674.
10. Kalles, D., & Pierrakeas, Ch. (2006). Using genetic algorithms and decision trees for a posteriori analysis and evaluation of tutoring practices based on student failure models. Proceedings of the 3rd IFIP conference on Artificial Intelligence Applications and Innovations, Athens, Greece, pp. 9–18, Berlin Heidelberg New York: Springer.
11. Hadzilacos, Th., & Kalles, D. (2006). On the software engineering aspects of educational intelligence. Proceedings of the 10th International Conference on Knowledge-Based Intelligent Information & Engineering Systems, Bournemouth, UK, LNCS 4252, pp. 1136–1143, Berlin Heidelberg New York: Springer.
12. Hadzilacos, Th., Kalles, D., Pierrakeas, Ch., & Xenos, M. (2006). On small data sets revealing big differences. Proceedings of the 4th Panhellenic Conference on Artificial Intelligence, Heraklion, Greece, LNCS 3955, pp. 512–515, Berlin Heidelberg New York: Springer.
13. Mitchell, T. (1997). Machine Learning. New York: McGraw Hill.
14. Quinlan, J.R (1993). C4.5: Programs for machine learning. San Mateo, CA: Morgan Kaufmann.

15. Breiman, L. (1996). Bagging predictors. Machine Learning, 24:2, 123–140.
16. Rumelhart, D., Widrow, B., & Lehr, M. (1994). The basic ideas in neural networks. Communications of the ACM, 37:3, 87–92.
17. Perlich, C., Provost, F. & Simonoff, J.S. (2004). Tree induction vs. logistic regression: a learning-curve analysis. Journal of Machine Learning Research, 4:2, 211–25.
18. Burges, C. (1998). A tutorial on support vector machines for pattern recognition. Data Mining and Knowledge Discovery, 2, 1–47.
19. Werth, L.H. (1986). Predicting student performance in a beginning computer science class. Proceedings of the 17th SIGCSE Technical Symposium on Computer Science Education, Cincinnati, OH, pp. 138–143.
20. Minaei-Bidogli, B., Kashy, D.A., Kortemeyer, G., & Punch, W.F. (2003). Predicting student performance: an application of data mining methods with the educational web-based system LON-CAPA. Proceedings of the 33rd ASEE/IEEE Frontiers in Education conference, Boulder, CO.
21. Ringwood, J.V., Devitt, F., Doherty, S., Farell, R., Lawlor, B., McLoone, S.C., McLoone, S.F., Rogers, A., Villing, R., & Ward, T. (2005). A resource management tool for implementing strategic direction in an academic department. Journal of Higher Education Policy and Management 27(2): 273–283.
22. Whittington, L.A. (1995). Factors Impacting on the Success of Distance Education Students of the University of the West Indies: A Review of the Literature. Review, University of West Indies.
23. Scott, D. (2005). Retention, completion and progression in tertiary education in New Zealand. Journal of Higher Education Policy and Management 27(1): 3–17.

9

Group Adaptation and Group Modelling

Judith Masthoff

University of Aberdeen, Aberdeen, Scotland, UK, jmasthof@csd.abdn.ac.uk

Summary. This chapter shows how a system can adapt to a group of users by aggregating information from individual user models and modelling the users' affective state. It summarizes results from previous research in this area. It also shows how group adaptation techniques can be applied when adapting to individuals, in particular for solving the cold-start problem and dealing with multiple criteria.

9.1 Introduction

Almost all work on adaptive systems to date focuses on adapting to *individual* users. For instance, a recommender system may select a book for a particular user to read based on a model of that user's preferences in the past. An intelligent tutoring system may adapt its instruction to an individual learner based on a model of that learner's interests and performance. The challenge adaptive system designers traditionally faced is how to decide what would be optimal for an individual user (so, what books would they like, what lesson is at the right level of difficulty). A lot of progress has been made on this (some of which is visible in other chapters in this book).

In this chapter, we go one-step further. There are many situations when it would be good if we could adapt to a *group* of users rather than to an individual. For instance, a recommender system may select a book for a reading group to read, based on models of all group members, and an intelligent tutoring system may adapt its instruction to a class. Adapting to groups is even more complicated than adapting to individuals. Assuming that we know perfectly what is good for individual users, the issue arises how to combine individual user models. In this chapter, we will discuss how group adaptation works, what its problems are, and what advances have been made. Interestingly, we will show that group adaptation techniques have many uses as well when adapting to individuals. So, even if you are developing systems aimed at individual users you may still want to read on (perhaps reading Sect. 9.7 first will convince you).

J. Masthoff: *Group Adaptation and Group Modelling*, Studies in Computational Intelligence (SCI) **104**, 157–173 (2008)
www.springerlink.com

In the next section, we will highlight some more scenarios in which group adaptation is needed. Section 9.3 discusses strategies for combining models of individual users to allow for group adaptation, and what we have learned from our experiments in this area. Section 9.4 deals with the issue of order when we want to recommend a sequence of items. Section 9.5 provides an introduction into the modelling of affective state, including how an individual's affective state can be influenced by the affective states of other group members. Section 9.6 explores how such a model of affective state can be used to build more sophisticated aggregation strategies. Section 9.7 shows how group modelling and group adaptation techniques can be used when adapting to an individual user. Section 9.8 provides an overview of other work on group adaptation and group modelling. Section 9.9 concludes this chapter.

9.2 Scenarios

There are many circumstances in which adaptation to a group is needed rather than to an individual. Below, we present the two scenarios that inspired our work in this area.

9.2.1 Interactive Television

Interactive television offers the possibility of personalized viewing experiences. For instance, instead of everybody watching the same news program, it could be personalized to the viewer. For me, this could mean adding more stories about the Netherlands (where I come from), China (a country that fascinates me after having spent some holidays there) and football, but removing stories about cricket (a sport I hardly understand) and local crime. Similarly, music programs could be adapted to show music clips that I actually like.

There are two main differences between traditional personalization as it applies to say PC-based software and the interactive TV scenarios sketched above. Firstly, in contrast to the use of PCs, television viewing is largely a family or social activity. So, instead of adapting the news to an individual viewer, the television would have to adapt it to the *group* of people sitting in front of it at that time. Secondly, traditional work on personalization has often concerned adapting one particular thing to the user, so for instance, recommending which movie the user should watch. In the scenarios sketched above, the television needs to adapt a *sequence* of items (news items, music clips) to the viewer. The combination of adapting to a group and adapting a sequence is very interesting, as it may allow you to keep all individuals in the group satisfied by compensating for items a particular user dislikes with other items in the sequence which they do like.

9.2.2 Ambient Intelligence

Ambient intelligence deals with designing physical environments that are sensitive and responsive to the presence of people. For instance, consider the case of a bookstore where sensors detect the presence of customers identified by some portable device (e.g. a Bluetooth-enabled mobile phone, or a fidelity card equipped with an active RFID tag). In this scenario, there are various sensors distributed among the shelves and sections of the bookstore which are able to detect the presence of individual customers. The bookstore can associate the identification of customers with their profiling information, such as preferences, buying patterns and so on.

With this infrastructure in place, the bookstore can provide customers with a responsive environment that would adapt to maximise their well-being with a view to increasing sales. For instance, the device playing the background music should take into account the preferences of the group of customers within hearing distance. Similarly, LCD displays scattered in the store show items based on the customers nearby, the lights on the shop's display window (showing new titles) can be rearranged to reflect the preferences and interests of the group of customers watching it, and so on. Clearly, group adaptation is needed, as most physical environments will be used by multiple people at the same time.

9.3 Aggregation Strategies

The main problem group adaptation needs to solve is how to adapt to the group as a whole based on information about individual users' likes and dislikes[1]. For instance, suppose the group contains three people, Peter, Jane and Mary. Suppose a system is aware that these three individuals are present and knows their interest in each of a set of items (e.g. music clips or advertisements). Table 9.1 gives example ratings on a scale of 1 (really hate) to 10 (really like). Which items should the system show, given time for four items?

Many different strategies exist for aggregating ratings of individuals into a rating of a group (e.g. used in elections, like when selecting the leader of a political party). Eleven of these (inspired by Social Choice Theory) are discussed in [1]. For instance, one could average the ratings of the individuals to obtain a group rating (making E and F the most preferred items by the group): the Average Strategy. One could take the minimum of the ratings, assuming that a group is as happy as its least happy member (see Table 9.2): the Least Misery Strategy.

One could use a combination of the Average and Least Misery strategy, taking the average of ratings but only for those items whose ratings are all

[1] Of course, there are other problems, like finding out what individual users like in the first place. We ignore this issue in this chapter, as traditional personalization techniques can be used for this.

Table 9.1. Example of individual ratings for ten items (A to J)

	A	B	C	D	E	F	G	H	I	J
Peter	10	4	3	6	10	9	6	8	10	8
Jane	1	9	8	9	7	9	6	9	3	8
Mary	10	5	2	7	9	8	5	6	7	6

Table 9.2. Example of the least misery strategy

	A	B	C	D	E	F	G	H	I	J
Peter	10	4	3	6	10	9	6	8	10	8
Jane	1	9	8	9	7	9	6	9	3	8
Mary	10	5	2	7	9	8	5	6	7	6
Group rating	1	4	2	6	7	8	5	6	3	6

Fig. 9.1. Experiment 1: which sequence of items do people select if given the system's task

above a threshold: The Average Without Misery Strategy. We conducted a series of experiments to investigate which strategy is best (see [2] for details).

In Experiment 1 (see Fig. 9.1), we investigated how people would solve this problem, so given ratings for individuals (as in Table 9.1), which items they thought the group should watch, if there was time for say six items. We compared our subjects' decisions (and rationale) with those of the aggregation strategies. We found that humans care about fairness, and about preventing misery and starvation ("this one is for Mary, as she has had nothing she liked so far"). Subjects' behaviour reflected that of several of the strategies (e.g. Average, Least Misery, and Average Without Misery were used), while other strategies were clearly not used.

In Experiment 2 (see Fig. 9.2), we presented subjects with item sequences chosen by the aggregation strategies. Subjects rated how satisfied they thought the group members would be with those sequences, and explained their ratings. We found that the Multiplicative Strategy (which multiplies the individual ratings) performed best, in the sense that all subjects thought its sequence would keep all members of the group satisfied. Several strategies

Fig. 9.2. Experiment 2: what do people like?

could be discarded as they clearly were judged to result in misery for group members. We also compared the subjects' judgements with predictions by simple satisfaction modelling functions. Amongst other, we found that more accurate predictions resulted from using:

- Quadratic ratings, which, e.g. makes the difference between a rating of 9 and 10 bigger than that between a rating of 5 and 6.
- Normalization, which takes into account that people rate in different ways, e.g., some always use the extremes of a scale, while some others only use the middle of the scale.

9.4 Impact of Sequence Order

To select a sequence of items, a system could use an aggregation strategy (like the Multiplicative Strategy) and then select the items with the highest ratings for the group. Of course, the question arises what order to show the items in, e.g., from highest rating to lowest or mixed. It seems likely that some kind of optimization is needed to get the best order.

However, the problem is actually far more complicated than that. Firstly, in responsive environments, the group membership changes continuously, so deciding on the next five items to show based on the current members seems not a sensible strategy, as in the worse case, none of these members may be present anymore when the fifth item is shown. Secondly, overall satisfaction with a sequence may depend on the order of the items: for instance, it may be good for satisfaction to have mood consistency (not putting a depressing item in the middle of two happy ones), have a strong ending, and provide a good narrative flow.

In Experiment 3 (see Fig. 9.3), we investigated how a previous item may influence the impact of the next item. Amongst others, we found that mood (resulting from the previous item) and topical relatedness can influence ratings for subsequent items. This means that aggregating individual profiles into a group profile should be done repeatedly, every time a decision needs to be made about the next item to display.

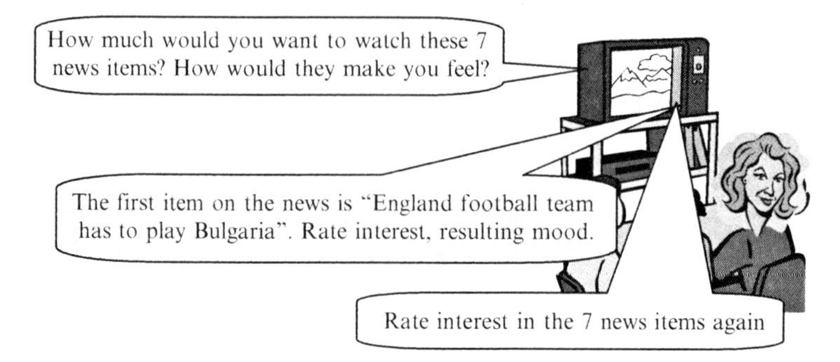

Fig. 9.3. Experiment 3: investigating the effect of mood and topic

9.5 Modelling Affective State

When adapting to a group of people, you cannot give everybody what they like all of the time. However, you do not want anybody to get too dissatisfied. For instance, in a shop it would be bad if a customer were to leave and never come back, because they really cannot stand the background music. Many shops currently opt to play music that nobody really hates, but most people not love either. This may prevent loosing customers, but would not result in increasing sales. An ideal shop would adapt the music to the customers in hearing range in such a way that they get songs they really like most of the time (increasing the likelihood of sales and returns to the shop). To achieve this, it is unavoidable that customers will occasionally get songs they hate, but this should happen at a moment when they can cope with it (e.g. when being in a good mood because they loved the previous songs). Therefore, it is important to monitor continuously how satisfied each group member is. Of course, it would put an unacceptable burden on the customers if they had to rate their satisfaction (on music, advertisements etc) all the time. Similarly, measuring this satisfaction via sensors (like heart rate monitors or facial expression recognizers) is not yet an option, as they tend to be too intrusive, inaccurate or expensive. So, we propose to model group members' satisfaction; predicting it based on what we know about their likes and dislikes.

In [3], we investigated four satisfaction functions to perform this modelling. We compared the predictions of these satisfaction functions with the predictions of real users. We also performed an experiment (see Fig. 9.4) to compare the predictions with the real feelings of users.[2] The satisfaction function that performed best defines the satisfaction of a user with a new item i after having seen a sequence *items* of items as:

$$\text{Sat}(items + < i >) = \frac{\delta \times \text{Sat}(items) + \text{Impact}(i, \delta \times \text{Sat}(items))}{1 + \delta}$$

[2] This was done in another (educational) domain. See [2] for a discussion of why this was necessary.

Fig. 9.4. Experiment 4: measuring overall satisfaction during a series of tasks

with the impact on satisfaction of new item i given existing satisfaction s defined as

$$\text{Impact}(i, s = \text{Impact}(i) + (s - \text{Impact}(i)) \times \varepsilon, \qquad \text{for } 0 \leq \varepsilon \leq 1 \text{ and } 0 \leq \delta \leq 1$$

Parameter δ represents satisfaction decaying over time (with $\delta = 0$ past items have no influence, with $\delta = 1$ there is no decay). Parameter ε represents the influence of the user's satisfaction based on previous items on the impact of a new item. The psychology and economics literature discussed in [3] shows that mood impacts evaluative judgement. For instance, half the subjects answering a questionnaire about their TVs received a small present first to put them in a good mood. These subjects were found to have televisions that performed better. Parameters δ and ε are user dependent (as confirmed in the experiment in [3]). We will not define Impact(i) in this chapter, see [3] for details, but it involves quadratic ratings and normalization as found in the experiment discussed above.

9.5.1 Effects of the Group on an Individual's Satisfaction

The satisfaction function given does not take the satisfaction of other users in the group into account, which may well influence a user's satisfaction. As argued in [3] (based on social psychology), two main processes can take place.

Emotional Contagion

Firstly, the satisfaction of other users can lead to so-called emotional contagion: other users being satisfied may increase a user's satisfaction (e.g. if somebody smiles at you, you may automatically smile back and feel better as a result). The opposite may also happen: other users being dissatisfied may decrease a user's satisfaction. For instance, if you are watching a film with a group of friends than the fact that your friends are clearly not enjoying it may negatively impact your own satisfaction.

Fig. 9.5. Types of relationship

Fig. 9.6. Experiment 5: impact of relationship type on emotional contagion

Emotional contagion may depend on your personality (some people are more easily contaged than others), and your relationship with the other person. Anthropologists and social psychologists have found substantial evidence for the existence of four basic types of relationships, see Fig. 9.5. In Experiment 5 (see Fig. 9.6), we confirmed that emotional contagion indeed depends on the relationship you have: you are more likely to be contaged by somebody you love (like your best friend) or respect (like your mother or boss) then by somebody you are on equal footing with or are in competition with.

Conformity

Secondly, the opinion of other users may influence your own expressed opinion, based on the so-called process of conformity.

Figure 9.7 shows the famous conformity experiment by Ash (see references in [3]). Subjects were given a very easy task to do, like decide which of the four lines has the same orientation as the line in Card A. They thought they were surrounded by other subjects, but in fact the others were part of the

Fig. 9.7. Conformity experiment by Ash

experiment team. The others all answered the question before them, picking the same wrong answer. It was shown that most subjects than pick that same wrong answer as well.

Two types of conformity exist: (1) normative influence, in which you want to be part of the group and express an opinion like the rest of the group even though inside you still belief differently, and (2) informational influence, in which your own opinion changes because you believe the group must be right. Informational influence would change your own satisfaction, while normative influence can change the satisfaction of others through emotional contagion because of the (insincere) emotions you are portraying.

More complicated satisfaction functions are presented in [3] to model emotional contagion and both types of conformity.

9.6 Using Affective State Inside Aggregation Strategies

Once you have an accurate model of the individual users' satisfaction, it would be nice to use this model to improve on the group aggregation strategies. For instance, the aggregation strategy could set out to please the least satisfied member of the group. This can be done in many different ways. We have only started to explore this issue. In [1], we describe some initial ways in which this can be done (as well as an agent-based architecture for applying these ideas to the ambient intelligent scenario and an implemented prototype).

Table 9.3. Results of average strategy with equal weights and with twice the weight for Jane

	A	B	C	D	E	F	G	H	I	J
Peter	10	4	3	6	10	9	6	8	10	8
Jane	1	9	8	9	7	9	6	9	3	8
Mary	10	5	2	7	9	8	5	6	7	6
Average (equal weights)	7	6	4.3	7.3	8.7	8.7	5.7	7.7	6.7	7.3
Average (Jane twice)	5.5	6.8	5.3	8.3	8.3	8.8	5.8	8	5.8	7.5

The Strongly Support Grumpiest strategy picks the item which is *most liked* by the least satisfied member. If multiple of these items exist, it uses one of the standard aggregation strategies, for instance the Multiplicative Strategy.

The Weakly Support Grumpiest strategy selects the items that are *quite liked* by the least satisfied member, for instance items with a rating of 8 or above. It uses one of the standard aggregation strategies, like the Multiplicative Strategy, to choose between these items.

An alternative approach would be to assign weights to users depending on their satisfaction, and then use a weighted form of a standard aggregation strategy. For instance, Table 9.3 shows the effect of assigning double the weight to Jane when using the Average Strategy. Note that weights are impossible to apply to a strategy like the Least Misery Strategy.

Clearly, empirical research is needed to investigate the best way of using affective state inside an aggregation strategy.

9.7 Applying Group Modelling to Individual Users

So, what if you are developing an application that adapts to a single user? Group adaptation techniques can be useful in three ways: (1) to aggregate multiple criteria, (2) to solve the so-called cold-start problem, (3) to take into account opinions of others.

9.7.1 Multiple Criteria

Sometimes it is difficult to give recommendations because the problem is multi-dimensional: multiple criteria play a role. For instance, in a news recommender system, a user may have a preference for location (being more interested in stories close to home, or related to their favourite holiday place). The user may also prefer more recent news, and have topical preferences (e.g. preferring news about politics to news about sport). The recommender system may end up with a situation like in Table 9.4, where different news story rate differently on the criteria. Which news stories should it now recommend?

Table 9.4 resembles the one we had for group adaptation above (Table 9.1), except that now instead of multiple users we have multiple criteria to satisfy. It is possible to apply our group adaptation techniques to this problem. However,

Table 9.4. Ratings on criteria for ten news items

	A	B	C	D	E	F	G	H	I	J
Topic	10	4	3	6	10	9	6	8	10	8
Location	1	9	8	9	7	9	6	9	3	8
Recency	10	5	2	7	9	8	5	6	7	6

there is an important difference between adapting to a group of people and adapting to a group of criteria. When adapting to a group of people, it seems sensible and morally correct to treat everybody equally. Of course, there may be some exceptions, for instance when the group contains adults as well as children, or when it is somebody's birthday. But in general, equality seems a good choice, and this was used in the group adaptation strategies discussed above. In contrast, when adapting to a group of criteria, there is no particular reason for assuming all criteria are as important. It is even quite likely that not all criteria are equally important to a particular person. Indeed, in an experiment we found that users treat criteria in different ways, giving more importance to some criteria (e.g. recency is seen as more important than location) [4]. So, how can we adapt the group adaptation strategies to deal with this? There are several ways in which this can be done:

- *Apply the strategy to the most respected criteria only.*
  The ratings of unimportant criteria are ignored completely. For instance, assume criterion Location is regarded unimportant, then its ratings are ignored. For instance, the result of the Average Strategy becomes:

	A	B	C	D	E	F	G	H	I	J
Topic	10	4	3	6	10	9	6	8	10	8
Recency	10	5	2	7	9	8	5	6	7	6
Group	20	9	5	13	19	17	11	14	17	14

Group List: AE(F,I)(H,J)DGBC

- *Apply the strategy to all criteria but use weights*
  The ratings of unimportant criteria are given less weight. For instance, in the Average Strategy, the weight of a criterion is multiplied with its ratings to produce new ratings. For instance, suppose criteria Topic and Recency were three times as important as criterion Location. The result of the Average Strategy becomes:

	A	B	C	D	E	F	G	H	I	J
Topic	30	12	9	18	30	27	18	24	30	24
Location	1	9	8	9	7	9	6	9	3	8
Recency	30	15	6	21	27	24	15	18	21	18
Group	61	36	23	48	64	60	39	51	54	50

Group List weight 3–1–3: EAFIHJDGBC weight 2–1–2: EFA(H,I)JDGBC

In case of the Multiplicative Strategy, multiplying the ratings with weights does not have any effect. In that strategy, it is better to use the weights as exponents, so replace the ratings by the ratings to the power of the weight. Note that in both strategies, a weight of 0 results in ignoring the ratings completely, as above.

- *Adapt a strategy to behave differently to important versus unimportant criteria: Unequal Average Without Misery*
  Misery is avoided for important criteria but not for unimportant ones. Assume criterion Location is again regarded as unimportant. The result of the Average Without Misery strategy with threshold 6 becomes

	A	B	C	D	E	F	G	H	I	J
Topic	10	4	3	6	10	9	6	8	10	8
Location	1	9	8	9	7	9	6	9	3	8
Recency	10	5	2	7	9	8	5	6	7	6
Group	21			22	26	26		23	20	22

Group List threshold 6: (EF)H(J,D)AI threshold 7: (EF)AI

We have some evidence that people's behaviour reflects the outcomes of these strategies [4], however, more research is clearly needed in this area to see which strategy is best. Also, more research is needed to establish when to regard a criterion as "unimportant".

9.7.2 Cold-Start Problem

A big problem for recommender systems is the so-called cold-start problem: to adapt to a user, the system needs to know what the user liked in the past. This is needed in content-based filtering to decide on items similar to the ones the user liked. It is needed in social filtering to decide on the users who resemble this user in the sense that they (dis)liked the same items in the past (see Fig. 9.8). So, what if you do not know anything about the user yet, because they only just started using the system? Recommender system designers tend to solve this problem by either getting users to rate items at the start, or by getting them to answer some demographic questions (and then using stereotypes as a starting point, e.g. elderly people like classical music).

Both methods require user effort. It is also not easy to decide which items to get a user to rate, and stereotypes can be quite wrong and offensive (some elderly people prefer pop music and people might not like being classified as elderly).

The group adaptation work presented in this chapter provides an alternative solution. When a user is new to the system, we simply provide recommendations to that new user that would keep the whole group of existing users happy. We assume that our user will resemble one of our existing users, though we do not know which one, and that by recommending something that would keep all of them happy, the new user will be happy as well.

Fig. 9.8. Cold-start problem in case of social-filtering

Fig. 9.9. Gradually learning about the user, and whom she resembles most

Gradually, we will learn about the new user's tastes, for instance, by them rating our recommended items or, more implicitly, by them spending time on the items or not. We provide recommendations to the new user that would keep the group of existing users happy *including* the new user (or more precisely, the person we now assume the new user to be). The weight attached to the new user will be low initially, as we do not know much about them yet, and will gradually increase. We also start to attach less weight to existing users whose taste now evidently differs from our new user.

Figure 9.9 shows an example of the adaptation: the system is including the observed tastes of the new user to some extent, and has started to reduce the weights of some of the other users. After prolonged use of the system, the user's inferred wishes will completely dominate the selection.

We have done a small–scale experiment using the MovieLens dataset to explore the effectiveness of this approach. Initial results are encouraging. More detail on the experiment and on applying group adaptation to solve the cold-start problem is given in [5].

9.7.3 Virtual Group Members

Finally, group adaptation can also be used when adapting to an individual by adding virtual members to the group. For instance, a parent may be fine with the television entertaining their child, but may also want the child occasionally to learn something. When the child is alone, the profile of the parent can be added to the group as a virtual group member, and the TV could try to satisfy both.

9.8 Other Work on Group Adaptation and Group Modelling

Only a few systems use group adaptation, in the way intended in this chapter. We will discuss most of them, namely MusicFX [6], PolyLens [7], Intrigue [8], the Travel Decision Forum [9], and work by Yu et al. [10].

MusicFX is used in a company's fitness centre to select background music to suit a group of people working out at a given time. It uses a more complex version of the Average Without Misery strategy. Users rate all music stations, from +2 (really love this music) to −2 (really hate this music). These ratings are converted to positive numbers (by adding 2) and then squared to widen the gap between popular and less popular stations. An Average Without Misery strategy is used to generate a group list. To avoid starvation and always picking the same station, a weighted random selection is made from the top m stations of the list (m being a system parameter).

PolyLens is a group recommender extension of MovieLens, which recommends movies based on an individual's taste as inferred from ratings and social filtering. It allows users to create groups and ask for recommendations for that group. PolyLens uses the Least Misery strategy, assuming groups of people going to watch a movie together tend to be small and a small group to be as happy as its least happy member.

Intrigue recommends places to visit for tourist groups taking into account characteristics of subgroups within that group (such as children and the disabled). It uses a weighted form of the Average strategy, with weights depending on the number of people in the subgroup and the subgroup's relevance (children and disabled were given a higher relevance).

The Travel Decision Forum helps a group of users to agree on the desired attributes of a planned joint vacation. Users indicate their preferences on a set of dimensions (like sport and room facilities). For each dimension, the system aggregates the preferences, and users interact with embodied conversational agents representing other group members to reach an accepted group preference.[3] The aggregation does not seem to use the idea of fairness, in the sense that loosing out on one dimension does not seem to be compensated by getting your way on another.[4] Emphasis is put on making the aggregation

[3] This differs from the multiple criteria problem, in that the system does not have to aggregate the dimensions, as it is not recommending holidays.

[4] Users can indicate how important a subset of dimensions (like health facilities) is to them, so this could in principle be used for weighting if trying to establish fairness.

strategy nonmanipulable, in the sense that users cannot steer the outcome to their advantage by deliberately giving, e.g., extreme ratings that do not truly reflect their opinions. It should be noted that strategies like the Least Misery strategy are manipulable, in the sense that devious users could avoid getting items they do not like, even if they dislike them only slightly, by giving extremely negative ratings. Normalization of ratings can be used to make such a strategy less manipulable. In the kind of scenarios we have discussed in this chapter, users would not normally be aware of the ratings given by others. Also, often users would not explicitly provide ratings, but instead such ratings would be inferred from user behaviour (and using content-based or collaborative filtering methods). Both of these aspects make manipulability less of a problem.

Yu et al. [10] have worked on recommending a TV program to a group of users by aggregating individual user models; we shall call their system Yu's TV Recommender. Their work differs from ours in the sense that they focus on recommending just one program, rather than a sequence of items. However, they look at multiple features of TV programs; so, instead of having one rating per user, they have a set of ratings per user, one for each feature. Actually, they use ratings of -1 (user dislikes the feature), $+1$ (user likes the feature) and 0. Aggregation is based on total distance minimization (of the group model's feature vector compared to the individual members' feature vectors). This keeps most users in the group happy. It should be noted that this does not prevent some users being miserable. Indeed, in an evaluation they found that their aggregation worked well when the group was quite homogenous, but the results were disliked when the group was quite heterogeneous. Of course, finding a program to recommend to a group where everybody has different tastes is difficult and may be impossible. When recommending a sequence (as discussed in this chapter), fairness and preventing one user from being miserable all the time is more important.

As mentioned above, most of the existing systems use one of the aggregation strategies we investigated (sometimes with a small variation), and they differ in the one selected. Though some exploratory evaluation of MusicFX, PolyLens, and the Travel Decision Forum has taken place, for none of these systems it has been investigated how effective their strategy really is, and what the effect would be of using a different strategy. The experiments presented in this chapter shed some light on this question. In contrast, some evaluation of Yu's TV Recommender has taken place [10].

The application domains of PolyLens, MusicFX, and Yu's TV Recommender differ from the scenarios described in this chapter in the sense that these systems do not need to select a group of items: people normally only see one movie per evening, music stations can play forever[5], and Yu's TV Recommender chooses one TV program only. For Intrigue, on the other hand,

[5] This would have been different if MusicFX selected individual songs rather than radio stations.

it is quite likely that a tourist group would visit multiple attractions during their trip, but the selection of a balanced sequence has not been addressed yet. None of the existing systems models the user's affective state.

The term "group modelling" is also used for work that is quite different from that presented in this chapter. A lot of work has been on modelling common knowledge between group members (e.g. [11, 12]), modelling how a group interacts (e.g. [13, 14]) and group formation based on individual models ([13, 15]).

9.9 Conclusions

Group adaptation is a relatively new research area. In this chapter, we have shown that adapting to groups is important in various scenarios. We have also shown that it is not only important when adapting to groups of people, but can also be applied when adapting to individuals, e.g. to prevent the cold-start problem and deal with multiple criteria. This chapter is intended as an introduction in the area. For more detail please see [1–5, 9].

One might think that accurate predictions of individual satisfaction can also be used to improve the transparency of adaptive systems: showing how satisfied others in your group are, or how satisfied criteria are, could improve the users' understanding of the working of the system and perhaps make it easier to accept items they do not like. This may be a good idea for a system that adapts to individual users. However, in a group adaptation system, users' need for privacy is likely to conflict with their need for transparency. An important task of a group adaptation system is to avoid embarrassment. Users often like to conform to the group to avoid being disliked (we discussed normative conformity as part of Sect. 9.5.1 on how others in the group can influence an individual's affective state). In [3], we have investigated how different group aggregation strategies may affect privacy. More work is needed on user interfaces for systems that adapt to groups, in particular on how to balance privacy with transparency and scrutability.

The work presented in this chapter is only a starting point. There are many directions for further research. For instance, more research is needed on the optimal order of item sequences, on the modelling of affective state, on the use of affective state within an aggregation strategy and on explanations of group recommendations. Empirical evaluations will be vital to bring this field forwards.

Acknowledgements

Judith Masthoff's research is partly supported by Nuffield Foundation Grant No. NAL/00258/G.

References

1. Masthoff, J., Vasconcelos, W.W., Aitken, C., Correa da Silva, F.S.: Agent-Based Group Modelling for Ambient Intelligence. AISB Symposium on Affective Smart Environments, Newcastle, UK (2007)
2. Masthoff, J.: Group Modeling: Selecting a Sequence of Television Items to Suit a Group of Viewers. UMUAI 14 (2004) 37–85
3. Masthoff, J., Gatt, A.: In Pursuit of Satisfaction and the Prevention of Embarrassment: Affective State in Group Recommender Systems. UMUAI 16 (2006) 281–319
4. Masthoff, J.: Selecting News to Suit a Group of Criteria: An Exploration. 4th Workshop on Personalization in Future TV – Methods, Technologies, Applications for Personalized TV, Eindhoven, The Netherlands (2004)
5. Masthoff, J.: Modeling the Multiple People That Are Me. In: P. Brusilovsky, A. Corbett, F. de Rosis (eds.). Proceedings of the 2003 User Modeling Conference, Johnstown, PA. Springer, Berlin Heidelberg New York (2003) 258–262
6. McCarthy, J., Anagnost, T.: MusicFX: An Arbiter of Group Preferences for Computer Supported Collaborative Workouts. CSCW, Seattle, WA (1998) 363–372
7. O'Conner, M., Cosley, D., Konstan, J.A., Riedl, J.: PolyLens: A Recommender System for Groups of Users. ECSCW, Bonn, Germany (2001) 199–218. As accessed on http://www.cs.umn.edu/Research/GroupLens/poly-camera-final.pdf
8. Ardissono, L., Goy, A., Petrone, G., Segnan, M., Torasso, P.: Tailoring the Recommendation of Tourist Information to Heterogeneous User Groups. In S. Reich, M. Tzagarakis, P. De Bra (eds.), Hypermedia: Openness, Structural Awareness, and Adaptivity, International Workshops OHS-7, SC-3, and AH-3. Lecture Notes in Computer Science 2266, Springer, Berlin Heidelberg New York (2002) 280–295
9. Jameson, A.: More than the Sum of Its Members: Challenges for Group Recommender Systems. International Working Conference on Advanced Visual Interfaces, Gallipoli, Italy (2004)
10. Zhiwen Yu, Z., Zhou, X., Hao, Y. Gu, J.: TV Program Recommendation for Multiple Viewers Based on User Profile Merging. UMUAI 16 (2006) 63–82
11. Introne, J., Alterman, R.: Using Shared Representations to Improve Coordination and Intent Inference. UMUAI 16 (2006) 249–280
12. Suebnukarn, S., Haddawy, P.: Modeling Individual and Collaborative Problem-Solving in Medical Problem-Based Learning. UMUAI 16 (2006) 211–248
13. Read, T., Barros, B., Bárcena, E., Pancorbo, J.: Coalescing Individual and Collaborative Learning to Model User Linguistic Competences. UMUAI 16 (2006) 349–376
14. Harrer, A., McLaren, B.M., Walker, E., Bollen L., Sewall, J.: Creating Cognitive Tutors for Collaborative Learning: Steps Toward Realization. UMUAI 16 (2006) 175–209
15. Alfonseca, E., Carro, R.M., Martín, E., Ortigosa, A., Paredes, P.: The Impact of Learning Styles on Student Grouping for Collaborative Learning: A Case Study, UMUAI 16 (2006) 377–401

10

PNS: A Personalized News Aggregator on the Web

Georgios Paliouras[1], Alexandros Mouzakidis[1], Vassileios Moustakas[1,2], and Christos Skourlas[2]

[1] Institute of Informatics and Telecommunications, National Center for Scientific Research "Demokritos", Athens, Greece, paliourg@iit.demokritos.gr, alexm@iit.demokritos.gr, bmoustakas@iit.demokritos.gr

[2] Department of Informatics, Technological Institute of Athens, Athens, Greece, cskourlas@teiath.gr

Summary. This paper presents a system that aggregates news from various electronic news publishers and distributors. The system collects news from HTML and RSS Web documents by using source-specific information extraction programs (wrappers) and parsers, organizes them according to pre-defined news categories and constructs personalized views via a Web-based interface. Adaptive personalization is performed, based on the individual user interaction, user similarities and statistical analysis of aggregate usage data by machine learning algorithms. In addition to the presentation of the basic system, we present here the results of a user study, indicating the merits of the system, as well as ways to improve it further.

10.1 Introduction

In recent years, the World Wide Web has experienced a self-feeding increase in the number of users and the quantity of content, data and services. More content makes the Web more interesting for more users, who in turn create more content. This spiral effect seems now to be accelerated by Web 2.0 technologies and the ever-increasing possibilities for user-generated media. A typical example of this is the news industry, which seems to be turning fully online and trying to follow the developments in Web publishing. Most of the news publishers have introduced electronic versions of their content, which in many cases are much richer in structure than the traditional paper versions. Additionally, a number of intermediate services have appeared, such as thematic news portals, which aggregate and re-distribute information from various sources.

In this manner, the end user has gained access to an enormous volume of information, which apart from its clear positive side brings along the problem of information overload. The task of finding interesting information in all that is within reach is as daunting and frustrating for the non-expert user as

G. Paliouras et al.: *PNS: A Personalized News Aggregator on the Web*, Studies in Computational Intelligence (SCI) **104**, 175–197 (2008)
www.springerlink.com

looking for a needle in a hay-stack. Thus, if we are to support this exciting development, we need to devise better and simpler methods of access to interesting information. Personalization is one way of achieving this through the modelling of user interests. Personalization systems typically acquire models of individual users or groups of users and then use these models to filter content, to recommend interesting content or to facilitate search. The acquisition of the user models is either done "manually", i.e. by asking the users or experts to define them, or in a less obtrusive automated manner, by statistical analysis of usage data.

PNS (Personalized News Service),[1] the system that we present here, attempts to personalize the experience of news reading at the level of an intermediate aggregating news service. PNS is a portal that aggregates news from various multi-language electronic news sources and provides a user with a personalized view of recent and past news items. Aggregation is done both through RSS feeds, as well as through information extraction with the help of simple HTML parsing programs (wrappers). Highlights of new items (title, source etc.) are retrieved periodically from various Web portals and e-newspapers. Once retrieved, they are organized according to predefined news categories and a Web-based interface provides personalized views to the users. Personalization in PNS is powered by a general-purpose personalization server called PServer. PServer uses statistical analysis and machine learning methods [14] to support four types of adaptive personalization: (a) personal user models, (b) user stereotypes, (c) user communities, (d) associated items. PNS uses all four types of personalization to rank news items according to the user's individual preferences or the interests of similar users. Thus, PNS is to our knowledge the only news aggregator supporting such powerful and flexible personalized news reading.

Compared to the previous version of PNS that was presented in [11], the version presented here is improved in several ways: (a) it provides a more complete personalization solution, integrating in a better way into the system various of the services available by PServer, (b) it includes many more news sources, which was noted as a major requirement of the users in the evaluation of the previous system, (c) the interface has improved significantly, and (d) a brand new user study has been performed.

The rest of this paper is structured as follows. Section 2 describes the design and implementation of PNS. Section 3 presents the results of an initial user study. Section 4 reviews the state-of-the-art systems for news personalization and in the last section conclusions and future directions are presented.

---

[1] http://pns.iit.demokritos.gr/

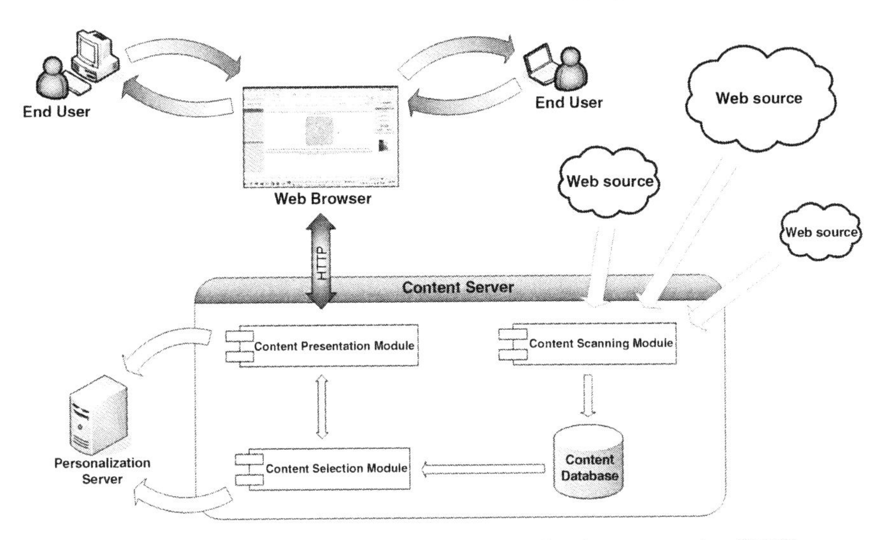

Fig. 10.1. The architecture of the personalized news service (PNS)

10.2 Personalized News Service

The Personalized News Service (PNS) provides its users with personalized access to news items harvested from multiple Web sources. It takes input from both the content sources (news agencies, news portals, electronic newspapers etc.) and the users themselves. This information is processed and a personal newspaper is constructed with recent news items that match the user's preferences. Figure 10.1 illustrates the system's overall architecture with emphasis on the Content Server which consists of the following basic modules: (a) the Content Scanner, (b) the Content Selector, (c) the Content Presenter, and (d) the Content Database, where information about about the news sources, the news items[2] and the wrappers are stored.

The system collects information about users in two ways:

- During the registration, the user specifies a user name and password and may also provide personal information, such as age, gender, occupation. Personal information is fed to the Personalization server (PServer) for improved personalization.
- During the use of the system, the users' browsing activity updates the corresponding user models maintained by PServer.

The component modules of the system and the basic functionality of PServer are described in more detail in the following sections.

[2] Respecting the copyright of the sources, the server does not store the content of the articles, but simply indexes it, according to its own categorization.

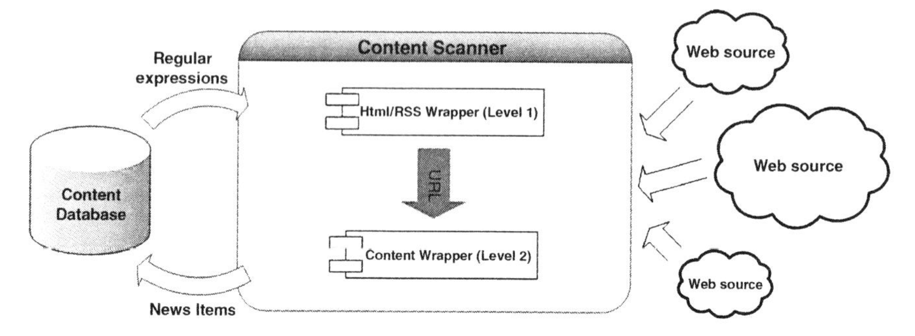

Fig. 10.2. The architecture of the content scanner

10.2.1 Content Aggregation

The content scanning module is responsible for locating and retrieving new items from a list of pre-specified sources and then storing in the database basic indexing information that will allow personalization and retrieval of the item. The aggregation process is done offline with a Web spider that is called periodically.

The spider works on the list of sources and associated URL addresses, which are stored in the database. For each source it follows a two-stage procedure: (a) identifying the addresses of new items in the source, (b) retrieving the items and extracting the information required for indexing them. For each of these two subprocesses an HTML wrapper is invoked, i.e., a small parsing module that identifies the required information within each Web page. Figure 10.2 illustrates this two-level identification and extraction process.

The first level of wrapping (HTML/RSS wrapper) involves the identification of URL addresses of new items in the source and per category of news. Thus, for each source-category pair an address is stored in the database and associated with a corresponding wrapper, which takes the form of a set of regular expressions. If the page that is retrieved is an RSS document, the wrapper has to parse the corresponding XML file and identify the URL addresses of new items and associated information. Typically, RSS documents, annotate the address of each time, each title and other information with XML tags. The information that we retrieve at the moment is the address and the title. For example an RSS document of a news source might have the following text:

```
<item>
    <title> Title of news item </title>
    <link> Address of news item </link>
    <pubDate> Publication date and time </pubDate>
    ...
</item>
```

If on the other hand, the source is not RSS, an HTML page is parsed by the wrapper and the addresses and titles of new items are extracted. The following is a sample extract of an HTML page containing links to articles:

```
<tr>
    <td align="justify" valign="top">
    <a class="title" href=" Address of news item" >
    <b> Title of news item </b></a>
    <br><div class="cat"> Publication date and time </div>
    ...
    </td>
</tr>
```

In order to build a wrapper in this case, one needs to identify the expressions that delimit the information of interest. For instance, the strings <a class= "title" href=" and " > delimit the address of the articles in the above example. In both RSS and HTML sources, new items are distinguished from old ones, based on their URL addresses.

The second level of wrapping (content wrapper) extracts useful information from each new item. This information is used for indexing and retrieving the article, as well as presenting a highlight of it to the user. So far, we are extracting only the first sentence of each article, parsing the corresponding HTML page. In future versions of the system, additional information, such as keywords from the content of the article will be extracted. The wrapper is constructed in a similar fashion as for the level 1 wrapper, i.e. by identifying sufficiently delimiting regular expressions. In some RSS feeds a short description of the article is provided. The description information is becoming increasingly common in RSS news sources. In that case, the second level wrapper is not needed for the information that we are currently extracting.

One problem with both types of wrapper for HTML pages, i.e. non-RSS sources, is that they are source-specific. Each source uses a different format for the presentation of articles and therefore we need different regular expressions for each one. However, almost all of these HTML pages are generated dynamically from a content database and therefore the same wrapper works for all news items. Furthermore, the format does not change very frequently (on average every few months) and therefore the wrappers require only occasional updating. Even that can be difficult sometimes though. For this reason, we are studying wrapper verification and wrapper induction methods [15] that will allow us to recognize when a wrapper has changed and automatically produce the correct wrapper for the new HTML format. Additionally, the increasing use of RSS feeds will eventually remove the need for source-specific wrappers.

In summary, the HTML/RSS wrapper reads from the database, a list of URL addresses corresponding to the source-category pairs and the associated regular expressions and produces as output a list of URL addresses for new items and their titles. This information is received by the content wrapper, together with the corresponding regular expressions and a full record for the new item, containing its address, title and first sentence, is stored in the content database.

10.2.2 Personalized Content Selection

News aggregation provides one-stop access to many sources, but at the same time reveals in a very immediate manner the problem of information overload. In other words, by combining information from many sources, the user becomes aware of the quantity of information out there and the difficulty of getting to the items of interest. As a result, dealing with the problem of overload is essential in the PNS.

More specifically, the system provides four different personalized views to the news items:

Personal news provides a content-based ranking of the interestingness of news items based on the personal model of the individual. For example, a user may prefer to read financial and sports news, while another might be interested specifically in the world news of yahoo.

Stereotype news provides collaborative ranking of the items based on the model of all individuals with similar characteristics, i.e., age, gender, etc. Such personal information is optionally provided by the users when they register or can be added later and it is used for assigning an individual to a stereotype.

Community news provides collaborative ranking without taking into account the personal characteristics of the user, combining instead the models of the communities in which the user has been assigned automatically, e.g. users that prefer financial and sports news.

Related news provides collaborative navigation, using cluster models of items, in order to associate news and recommend further reading.

The four views are complementary in several dimensions. For instance, new users are likely to find the "stereotype news" more useful, as their personal model will be very poor. On the other hand, "related news" provides navigational help, in contrast to the ranking approach of the other three views. "Community news" is useful for expanding one's interests in a focussed way, especially when no personal information for the user is available, in order to assign the user to a stereotype.

Personalization is achieved with the use of a separate personalization server, called PServer,[3] that provides a variety of services. Pserver is a general-purpose personalization server that can be adapted to any kind of application requiring personalization services. Pserver works like a Web service, taking a request through the http protocol and re-turning XML documents with the

---

[3] PServer has been developed in the Institute of Informatics and Telecommunications of NCSR "Demokritos" and will soon be made available under a BSD-like license.

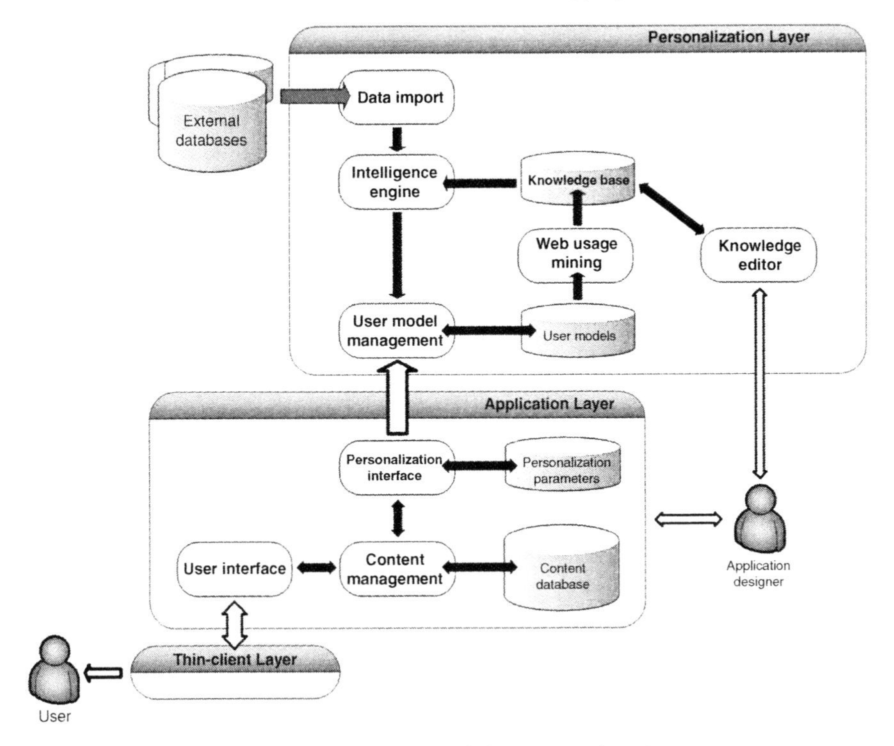

Fig. 10.3. The architecture of the personalization server

results, and can be used by many different applications concurrently. The developers who want to personalize their application do not need to make great modifications to their applications, but just the code required for making their application a client of Pserver. Thus, PServer makes the personalization of existing applications very easy. Figure 10.3 provides a high-level view of the PServer architecture.

The application layer in Fig. 10.3 illustrates a typical personalized application, such as PNS, while the Personalization server presents the main elements of the PServer. The user model management module is the one controlling the changes that happen to all types of user model. The Web usage mining process discovers new knowledge from the existing user models, which is then processed by the intelligence engine and provided to the management module, in order to update the user models. The optional introduction of external data or knowledge to the system, by data import and knowledge editing is not used in PNS. At the application layer, a typical personalized application, such as PNS, maintains a content database, selecting the content appropriate for each user. This content selection module communicates with the Personalization server, retrieving user models and updating them according to the users' actions.

The communication of PServer with the application is based on a common vocabulary of personalization parameters, which are defined by the application and communicated to PServer at the set-up stage of the system. There are two basic types of parameter, called *attributes* and *features*. Attributes capture information that is rather static, i.e., it is not processed statistically, but provided explicitly by the application. An example of such information is the personal information of each user, i.e., age, gender, etc. Features on the other hand capture usually the preferences of the users. In the case of PNS, the features correspond currently to news sources, news categories and their combination. As a result, personalization is only done at the coarse level of news sources and categories. Current work is expanding this to the contents of individual news items and clusters thereof. All users modeled by PServer share the same parameters, but not the same parameter values. PServer provides functions to insert of remove features, attributes and users, to get or set the values of the features for a specific user and functions to increase or to decrease the values of the features.

As explained above, PNS makes use of four types of adaptive personalization, using the corresponding services of PServer: (a) personal user models, (b) stereotype models, (c) community models, and (d) item clusters. Each of the four types requires the acquisition and maintenance of a different user model, which is achieved with the use of statistical analysis and machine learning methods. The corresponding services of Pserver are described below.

Personal user models: Each personal user model stores the attribute and feature values of an individual user. Features are updated according to the actions of each user, either as frequency counts and/or as a histories of actions. In this manner, we can at any point in time infer the level of interestingness of each user in a certain feature, such as the sports news in yahoo. In PNS, we are currently using only frequency counts.

Stereotype models: User stereotypes are sets of users with common attributes. For example all the users within a specific age range and a particular job type may constitute a stereotype. Like personal user models, stereotypes also have features that are updated according to the preferences of the users in the stereotype. However, in contrast to personal models, each stereotype may have a different feature set from all other stereotypes and thus each stereotype can be handled separately by the application. More advanced methods for learning stereotypes from personal models, e.g. [10], are not used currently in PNS.

Community models: The main problem with stereotypes is that users may provide inaccurate personal information, due to privacy or other concerns. For this reason, providing personal information is only an option in PNS. For those users who do not provide personal information, we need a different way to support collaborative filtering or ranking. This is achieved by the clustering of users into user communities. Communities in PServer are constructed on the basis of similarities between the users, using the cluster mining unsupervised learning algorithm. Any clustering method could be used for this purpose [12], but we are using the cluster mining method described in [13], due to the fact that it allows communities to overlap, i.e., each user may belong to more than one communities. The algorithm is a graph-based clustering method, associating communities with cliques in the graph of users. PServer is easily extensible with new algorithms for community discovery. This is done through a simple SDK and does not require the recompilation of the code of PServer.

Item clusters: In order to discover associations between features, e.g. news categories, one can simply cluster features together, according to their statistics in individual user models. This can be considered as the inverse task of community discovery. Hence, instead of clustering users, based on features, we cluster features based on users. In PServer, this is also achieved with the cluster mining algorithm, although other association discovery methods can easily be added.

10.2.3 Reading Personalized News

The content presenter module of PNS is basically the personalized graphical user interface of the system. The module is responsible for identifying the user, providing the various personalized views of PNS, informing PServer about the actions of the user and providing more traditional retrieval facilities, such as search by date or by keyword. In the following we illustrate the basic functionality of the module with the help of corresponding screenshots.

Entering the system, the user views a welcome screen, as shown in Fig. 10.4. From this point on the user can either browse all the news using the category and source menus on the left hand side of the screen or log into the system as

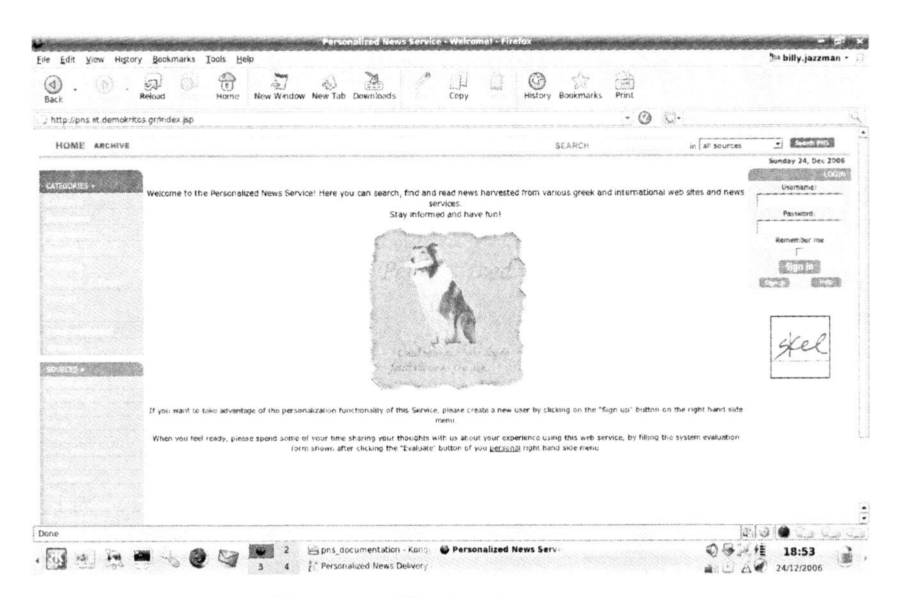

Fig. 10.4. The introductory page

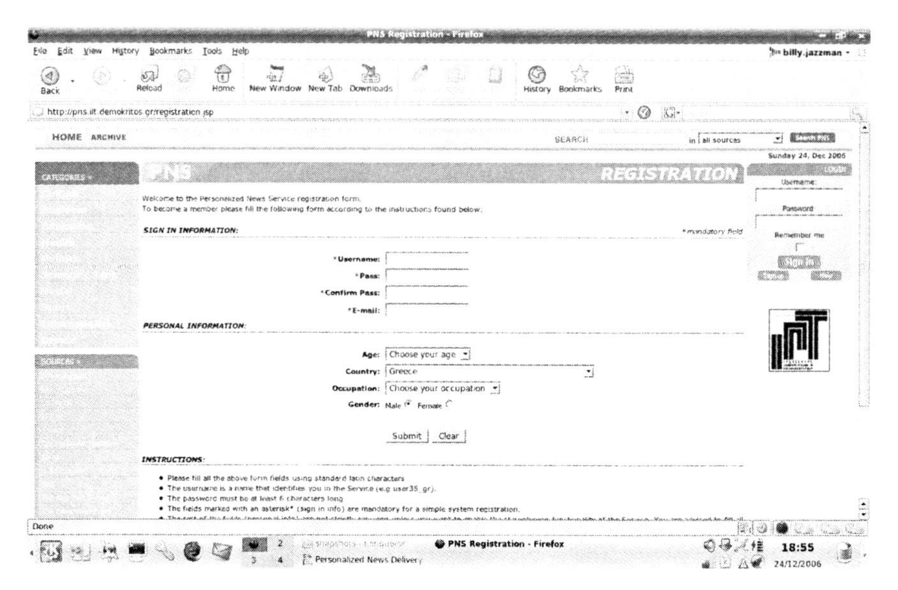

Fig. 10.5. The registration form

a known user (top right corner of the screen). If the user is not logged in, the news are presented in a non-personalized manner and no information is kept about the user's actions. By logging in, the system gets into a personalized mode and the news are presented according to the user's model. In particular the news are ranked according to their expected interestingness for the user, based on the personalization parameters that are used.

If the user is not already registered and wants to do so, they select to sign up and the registration form shown in Fig. 10.5 appears. The form is separated into a set of compulsory fields required for user identification, i.e., username, password and email address, and a set of optional fields that are to be used for assigning the user to a stereotype. If the user chooses not to provide these, stereotype-based personalization will not be available. A broader consequence of many users choosing not to provide personal information is that the stereotype models become statistically weak, as they are based only on a small fraction of the users who actually belong to the stereotype.

When logging into the system, the user moves to the "personal news" view described in Sect. 10.2.2 above. An example is shown in Fig. 10.6. In this view, the ranking of the articles is based on the information recorded in the user's profile, assumed to represent the user's preferences. Some basic information is shown for each article, avoiding the reproduction of the full article, due to copyright issues. In particular, only the title and the first sentence of the article are shown and a link to the original source is provided if the user wishes to read the whole article.

When a new user enters the system, this personalized view is actually the default one, as no information is available yet about the preferences of the

Fig. 10.6. The personal news page

user. In that case, the user can switch to the "stereotype news" view, using the menu on the top right corner of the screen. As explained above, this is only possible if the user has provided personal information. The effect on the ranking of the news items is similar to that of the "personal news" view, but a different model is used, i.e. the stereotype. The "community news" is also not meaningful for new users, as no information is available for assigning the user to communities of "similar" users. However, once the user has used the system a few times, assignment to communities becomes possible and the user can choose to rank the articles according to the models of the corresponding communities. The presentation of the news is again similar to that shown in Fig. 10.6.

The fourth personalization view discussed in Sect. 10.2.2, i.e. "related news" becomes available when the user selects to view an article. By selecting the article, a separate screen appears showing the article as shown in its original source, and adding an interesting link "Users who read this article also read ...", see Fig. 10.7. The PNS header is also added in order to show to the user that the system is still in a personalized mode.

When choosing the "Users who read this article also read ..." link, the user gets a list of articles ordered according to the clusters in which the first article belongs. The appearance of the ranked items is shown in Fig. 10.8 and is similar to that in Fig. 10.6. The user can then choose one of the recommended articles, causing its display, follow the "related news" link and so on. In this manner the navigation of the user in the content database becomes personalized in a collaborative way.

Fig. 10.7. Reading an article

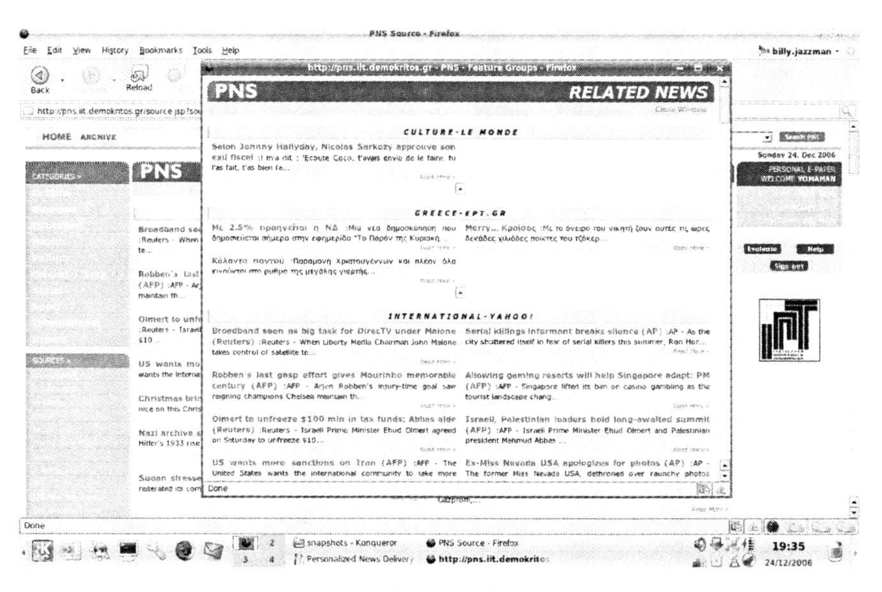

Fig. 10.8. Viewing related articles

Finally, in addition to the various personalized views, which refer to the most recent articles published in the corresponding sources, the user is able to search the content database by date, category and source (Fig. 10.9). By doing this, the user will retrieve all news published in the specified date range, and

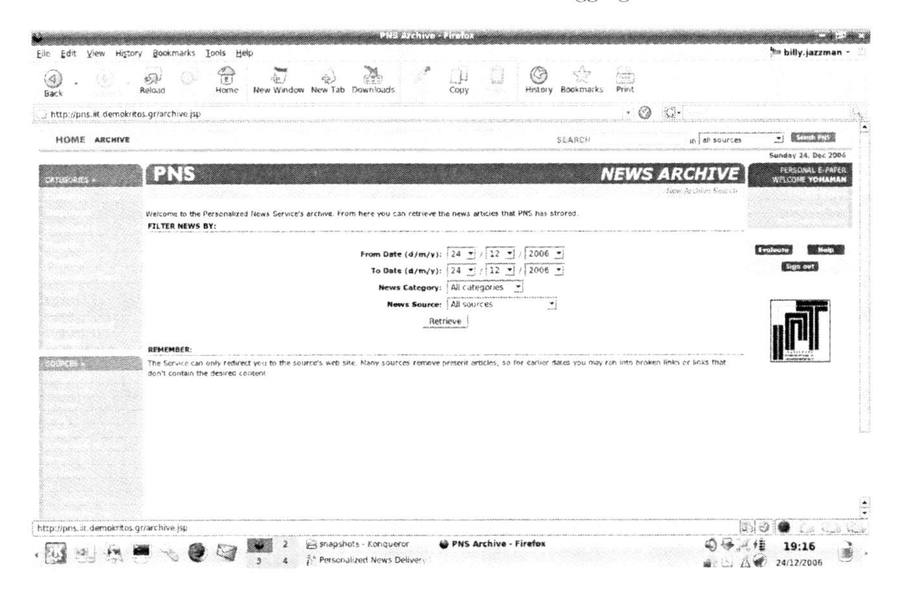

Fig. 10.9. The search-by-date page

in a particular source-category pair. In a similar manner, the user can search by a keyphrase within a particular source or in the whole database.

10.3 User Evaluation

10.3.1 Set-Up of the Study

In order to assess the usefulness of the system, users were asked to test the system for a short period of time. On a daily basis, the system collected the most recent news, which were then presented to the users. At the end of the test period, the users were asked to fill an electronic questionnaire with their observations and comments. The role of the user study was to gather feedback on several different aspects of PNS:

- Validating the personalization services
- Evaluating the functionality of the system
- Providing input to the design of the system

Thus, the questionnaire was separated into four sections. The first section asked for some basic characteristics of the evaluator, focusing mainly on computer literacy and use. The second section assessed the functionality of the system, focusing on usability issues. The third section, which is more interesting for this paper, assessed the value of different personalization views. Finally, the fourth section asked the users to provide suggestions for improvement.

In order to make the completion of the questionnaire easier and allow useful conclusions to be drawn even with a small number of users, most of the questions had a three-choice answer, like "Satisfied–Partially satisfied–Not satisfied" or "Very useful–Not so useful–Not useful".

At the end of the study we collected 34 answers and most of the users were highly literate in computer usage. Actually, the majority were either computer science students or academics. Therefore, the results that are presented below cannot be considered representative of the average user, but more biased on the technical issues of the system. This is particularly helpful for improving the system technically, but it is clear that a wider-audience evaluation is still needed.

10.3.2 Evaluating the Functionality of the System

In order to evaluate the usability of the system, the users were asked to respond to a number of questions concerning the user interface and important design parameters of the system, such as the news sources and news categories that are used. Most of these questions, though not all, had been asked also in the user study that was performed for the previous version of the system, presented in [11]. Table 10.1 presents the response of the users in the relevant questions and where available the results of the previous study are presented for comparison purposes.

The results of the study, regarding the functionality of the system are particularly encouraging. Comparing to the previous version of the system,

Table 10.1. Results on the functionality of the system

Do you find the web interface usable and comprehensible?			
Answer	Yes (%)	Partially (%)	No (%)
New study	73.5	23.5	2.9
Old study	70	30	0
How much time did it take you to get familiarized with the system?			
Answer	<30 min	About 60 min	>60 min
New study	76.5%	23.5%	0.0%
Did you find the news you are interested in quickly and easily?			
Answer	Yes (%)	Partially (%)	No (%)
New study	79.4	20.6	0.0
Old study	70	20	10
Are you satisfied with the news categories used by the system?			
Answer	Yes (%)	Partially (%)	No (%)
New study	55.9	44.1	0.0
Old study	20	45	35
Are you satisfied with the news sources included in the Service?			
Answer	Yes (%)	Partially (%)	No (%)
New study	41.2	58.8	0.0
Old study	10	50	40

usability has improved in all aspects. In the first three questions, there is a great majority, between 70 and 80% of people who are satisfied with various aspects of the user interface. Thus, there seems little room for improvement in this direction. Even more encouraging though are the results in the last two questions, regarding the news sources and categories that we are using. This aspect of the system has been criticized in the previous survey and the system seems to have improved significantly, since the answers have shifted towards the "satisfied" side of the spectrum, leaving the non-satisfied quadrant completely empty. Having said that, there is still room for improvement.

10.3.3 Evaluating the Personalization Aspects

The most interesting aspect of the study concerned the personalization functionality of PNS. For the first time, this study tried to assess the value of the individual personalization views, in addition to the overall value of personalization in PNS. Figure 10.10 shows the results that we obtained for the latter, which are almost identical with these obtained in the previous survey. This lack of improvement shows that we have not been able to provide the added value that we wanted to the user, through the use of the advanced personalization functionalities.

In an attempt to understand better where the problem lies, we asked more specific questions about the satisfaction of the users with the various personalization views. Figure 10.11 shows the results that we obtained. Based on these results, it seems that the problem focuses mainly on the collaborative views, as the personal news view is assessed rather positively. Of the collaborative views, "community news" is assessed very negatively and "related news" rather positively, while "stereotype news" is in between the two.

As an initial explanation of the situation, one needs to note the difficulty of assessing collaborative personalization. Especially communities require the use of the system for a significant amount of time and by a substantial number of people, in order to start adding value to the user. Stereotypes are a

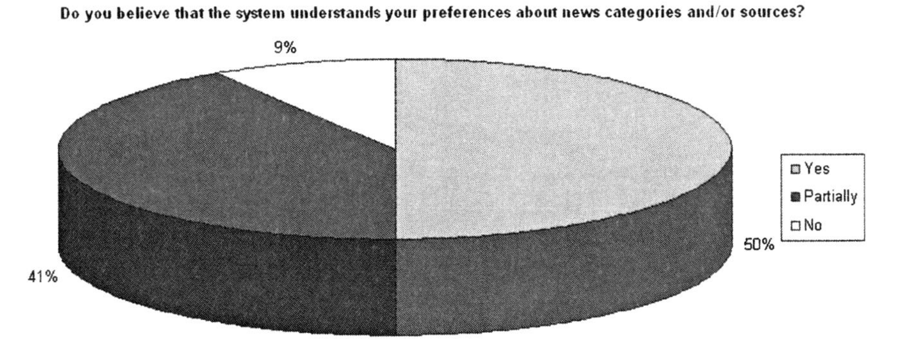

Fig. 10.10. Overall evaluation results for the personalization in PNS

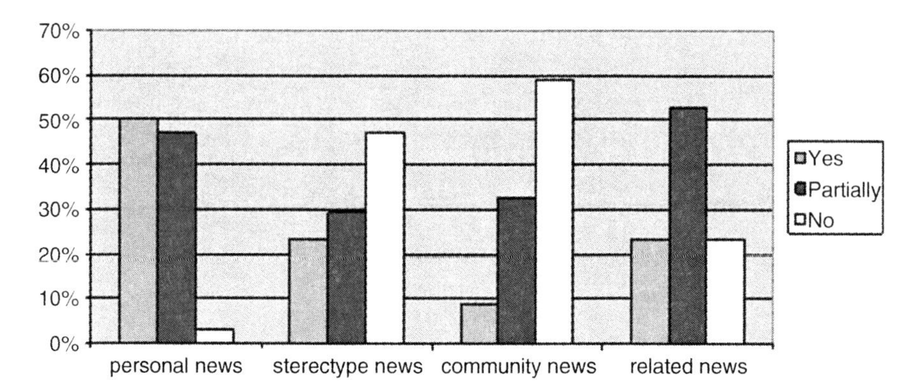

Fig. 10.11. Evaluation results for the various personalization views in PNS

Table 10.2. Correlation of responses for the collaborative views

	Stereotype news (%)	Community news (%)	Related news (%)
Stereotype news	100	63	81
Community news	63	100	58
Related news	81	58	100

bit better due to the fixed specification of the groups of people belonging to each stereotype, i.e., based on their personal characteristics which are entered at registration time. Still, though in order to obtain statistically significant evidence about the interests of each group more usage data are required than what we have been able to collect. Finally, in the "related news" view the situation is better, due to the small number of items, i.e., news sources and categories, that also have rather straightforward associations, e.g. when viewing a news item about politics, one would expect other politics news to be recommended.

Focusing further on the explanation of the problem, we measured the correlation in the users' responses, regarding the collaborative personalization views. Table 10.2 shows the percentage of common responses received for the three views. Each cell in the table shows the percentage of identical response for two views. For example, in 81% of the responses users gave the same assessment for "stereotype news" and "related news".

The above results show a low level of agreement in the assessment of "community news" with the other two views, which when combined with the negative assessment of this view, leads to the conclusion that agreement is mainly towards the negative responses, i.e., users who agree that the two views are equally unsatisfactory. On the other hand, there seems more positive agreement on the "stereotype news" and "related news" views.

10.3.4 Recommendations for Improvement

In addition to the quantitative evaluation of various aspects of the system, the users provided very interesting suggestions on how the system could be improved. Some of these concerned the functionality and the user interface of the system, while others addressed the personalization aspects of it.

Regarding the functionality of PNS, despite the positive quantitative results, we received some interesting suggestions to improve the system further. First, there was a common demand of the users to improve the way that articles are shown on the screen. The highly textual presentation seems tiring and a more visual approach is needed. There were also some suggestions about news sources from different countries and new categories or sub-categories, for example specialized technology news and news about specific kinds of music. These suggestions seem particularly relevant for the biased user group that took part in the evaluation, but a wider audience is expected to make different requests. Finally, we need to add a procedure for password recovery, as users often forget it.

On the personalization aspects of the system, the most common request was for more personalized suggestions. This is due to the fact that personalization is currently done only at the level of news sources and news categories. This is an important problem, which is also responsible to some extent for the negative assessment of collaborative personalization views. As a response to this problem, we are already extending personalization to the level of individual articles and their content, extending it also to the results of free-text queries on the archive. Additionally, users suggested that some of the collaborative personalization views, particularly communities, are an overkill for the system. As mentioned above, we believe that this is due to the short evaluation period and small number of users taking part, which did not allow the discovery of interesting communities. As a response to this comment we are looking into a tighter and more intuitive integration of the various collaborative personalization views, as well as the testing of new clustering methods. Furthermore, the evaluation of the next version of the system should be done over a longer period of time and with more users if possible. The initialization of the system with communities that have been discovered in previous user studies may also help in arriving more quickly at interesting community models.

10.4 Related Work

There exist many Web sites/portals available online that provide similar services to PNS. Many of them are experimental systems that were developed for research purposes while others are real-world commercial services. The category of commercial systems has been growing in the past couple of years as the potential of the technology has been acknowledged.[4] It is therefore important to see other systems' choices and directions to various problems in order to better place PNS on the map of existing systems.

In the next few paragraphs there is a small description of the philosophy and the various techniques and technologies incorporated in the most influential systems on the Web. The presentation is based on the two primary technologies adopted by PNS: first the aggregation feature is examined and then we look at how each system (if available) personalizes the provided information.

Content aggregation is mainly achieved nowadays through the use of RSS feeds. Feeds can be imported either statically or dynamically to fit a specific user's preferences. In the first case, this means that a set of news Web sites are imported by the system's administrator into the aggregation module and these constitute a global source pool from where all users draw information. Web services that use this static kind of aggregation include Findory,[5] EMM's News Brief[6] and NewsExplorer,[7] Cebil[8] and Phigita News.[9] Google News[10] takes this approach a step further as it facilitates the addition of new RSS sources through a public procedure were users ask for and evaluate sources to be included or even excluded from the system, as in the National Vanguard magazine controversy. On the other hand, other Web sites, give each individual user the ability to select their own sources and therefore be exempted from content they consider unreliable or of no interest to them. My Yahoo (http://my.yahoo.com) contributed significantly to the wide adoption of RSS, even by newcomers, by letting people customize their own My Yahoo page with feeds from any source. To that helped the acquisition of SearchFox a pioneering company in the area. Additionally a suggestion system using popularity or editorial picks informs new users for the "hottest" feeds available. A similar approach is followed by Netvibes,[11] topix[12] and Feeds2.0.[13]

News portals pioneered the development of personalized systems which adapt to a user's specific needs. The adaptation may concern something so primitive as the medium in which news are delivered to a variety of other things, such as the advertisements displayed.

[4] http://www.jdlasica.com/articles/personalization.html
[5] http://www.findory.com/
[6] http://emm.jrc.it
[7] http://press.jrc.it/NewsExplorer/home/en/latest.html
[8] http://www.cebil.gr/
[9] http://news.phigita.net/
[10] http://news.google.com/
[11] http://www.netvibes.com/
[12] http://www.topix.net/
[13] http://www.feeds2.com/

A basic ingredient to a successful personalization system is the way in which it collects user data and models each user. Findory, Spotback[14] and Feeds2.0 collect only implicit data from the user as the user navigates through the Web pages. This means that no registration process is needed, other than the creation and use of a user name and password. The user modelling is for the same reason adaptive as the model changes based on the analysis of usage data. Google News also tracks the browsing actions of users in order to collect implicit data. Additionally it requires full registration where the user provides data explicitly to the system such as demographic and preference information. All information is stored inside the user's Google Account, thus making the access of the personalized edition of Google News possible from any computer. A totally different approach is used by, the still in private beta, Leaptag.[15] LeapTag allows the user to define the things he is interested in using tagging. For each interest he/she creates a tag for, LeapTag will produce results that include news, blogs, books, etc. Advertisements are also personalized to the user's interest.

Finally, different personalized systems use different approaches to filtering and/or ranking of articles. These approaches fall into two basic categories, that have been briefly mentioned in previous sections: Content-based and collaborative filtering. Content-based filtering is based on the analysis of the article's content, aiming to identify important keywords for the user. The personal user models are an example of this type of filtering. On the other hand, collaborative filtering, such as stereotypes, communities and item clusters in PNS, is based on the assumption that users who regularly view the same articles have similar interests. Based on this assumption, recommendation mechanisms are built in order to help users to implicitly help each other find interesting articles.

There are several systems that provide content-based filtering. Examples are the Personal Wall Street Journal and the electronic edition of the San Francisco Chronicle which uses Fishwrap [6]. Other examples are: WebMate [5], the Mercurio system [7] of personalized access to the electronic variant of the Spanish ABC newspaper, NewsDude [4] and SmartPush [9]. On the other hand, an example of a collaborative filtering system is Findory.

Some web sites combine the two filtering approaches. Google News, Feeds2.0, Krakatoa [3] and its successor Anatagonomy [8] are some of them. Other such systems include SeAN [2] and the automated personalization system studied by Aggarwal and Yu [1]. Also Feeds 2.0 uses a language identification mechanism that allows the service to identify the language in which an item is written and automatically extract the most important keywords as tags for each item. Thus, search of items related to a particular object is made easier, while users are able to provide their own personal tagging of items.

---

[14] http://spotback.com/

[15] http://leaptag.com/

Table 10.3. Personalization techniques in research prototypes

	Data collection	User modelling	Filtering type
Fishwrap	Explicit	Non-adaptive	Content-based
Krakatoa, Anatagonomy	Explicit, implicit	Adaptive	Content-based, collaborative
SmartPush	Explicit	Non-adaptive	Content-based
SeAN	Explicit, implicit	Adaptive	Content-based, collaborative
Aggarwal and Yu	Explicit, implicit	Adaptive	Content-based, collaborative
WebMate	Explicit, implicit	Adaptive	Content-based
Mercurio (ABC newspaper)	Explicit	Adaptive	Content-based
NewsDude	Explicit	Adaptive	Content-based

Table 10.4. Personalization techniques in commercial systems

	Data collection	User modelling	Filtering type
Google News	Explicit, implicit	Adaptive	Content-based, collaborative
Findory	Implicit	Adaptive	Collaborative
Feeds2.0	Implicit	Adaptive	Content-based, collaborative
Personal Wall Street Journal	Explicit	Non-adaptive	Content-based
San Francisco Chronicle	Explicit	Non-adaptive	Content-based

Tables 10.3 and 10.4 summarize the personalization characteristics of the systems presented in this section.

PNS has two main differences from the systems presented above: the combination of RSS feeds with plain HTML sources and the provision of many complementary personalization views on the same data. The combination of RSS documents with HTML ones remains important for as long as there are interesting sources that do not use RSS feeds. The number of such sources is decreasing, but they still are the majority. It is expected that this situation will change in the medium to long term. The combination of multiple personalization views, powered by the use of PServer is more important. In its current state, PNS is able to personalize the reading of even new users, through stereotype modelling, while the added value increases for long-term users, both in the retrieval of articles, through personalized ranking, as well as in the navigation through the articles, using the "related news" view. The combination of content-based ("personal news") and collaborative ("community news") filtering is also particularly useful for a complete personalization solution.

10.5 Conclusions and Future Work

In this paper we have presented the Personalized News Service (PNS), which aggregates news from various Web sources and provides personalized access to it, using a variety of personalized views, namely "personal news", "stereotype news", "community news" and "related news". Personalization is powered by a general-purpose personalization server (PServer), which provides a variety of personalization capabilities to applications that require personalization. The use of this powerful personalization server is one of the distinguishing characteristics of PNS, allowing it to provide a complete personalization solution by integrating complementary types of personalization. Another characteristic of PNS, as compared to the state-of-the-art systems is its ability to process sources that do not provide their content in a structured format, i.e. through RSS feeds.

As part of this work, we have performed a user study, assessing various aspects of the system, and compared the results to those obtained for a previous version of the system. The results of the study have been particularly interesting, showing where more work is needed and also providing suggestions for improvement. Compared to the older version of the system, the satisfaction of the users with the basic functionality of the system and the user interface has increased significantly. Still it can be improved in several ways, such as the presentation of the articles that could become more user-friendly. Additionally, we would like to make the user interface more multilingual, as at the moment only English and Greek are supported.

Other improvements that we are working on are related to the maintenance of the wrappers for non-RSS sources, and the clustering of articles that talk about the same subject. Manual wrapper maintenance, i.e., changing the wrappers when the sources change their format, is becoming a major obstacle to the scalability of the system to many sources. Thus, we are integrating methods that we have been developing independently [15] to learn new wrappers without the intervention of the user. Article clustering is also becoming a major requirement as the number of sources increases, because the same article appears often in many sources. If these alternative versions of the article are treated as separate articles, the list of articles recommended by the system is going to become very large and with a high degree of repeated information.

The results of the user study have been particularly critical of collaborative personalization views, while the "personal news" view has been judged more positively. The users did not see the value of collaborative recommendations for the coarse level of personalization offered by the system, i.e. based on news sources and categories. Another reason for the criticism was the short evaluation period that did not allow for meaningful community models to be learned. Thus, one of our main goals is to move to a finer level of personalization, by using as personalization parameters the contents of individual articles. Furthermore, the evaluation of the next version of the system should be done over a longer period of time and if possible with a larger and more diverse user group.

In summary, apart from a very useful system for the end user, as a research prototype PNS has raised a number of interesting issues that we are trying to address with related methods that come out of our relevant research.

Acknowledgments

We would like to thank particularly the people who took part in the evaluation study. Their feedback is invaluable to us and we promise to do our best to improve the system in the directions that these comments point at. The presented work is part of a long-term project of the Software and Knowledge Engineering Laboratory at the Institute of Informatics and Telecommunications of NCSR "Demokritos". Part of this work was done in collaboration with the Department of Informatics of the Technological Institute of Athens, in the context of the research project PA_CO_CLIR (Parallel, Content Based Cross Language Information Retrieval) that is co-funded by the European Social Fund and National Resources (EPEAEK-II)-ARXIMHDHS.

References

1. C.C. Aggarwal and P.S. Yu. An automated system for web portal personalization. In *Proceedings of 28th International Conference on Very Large Data Bases (VLDB 2002)*, pages 1031–1040, 2002.
2. L. Ardissono, L. Console, and I. Torre. An adaptive system for the personalized access to news. *AI Communications*, 14(3):129–147, 2001.
3. K. Bharat, T. Kamba, and M. Albers. Personalized, interactive news on the web. *Multimedia Systems*, 6(5):349–358, 1998.
4. D. Billsus and M.J. Pazzani. A hybrid user model for news classification. In *Proceedings of the International Conference on User Modeling (UM), CISM Courses and Lectures, n. 407*, pages 99–108, 1999.
5. L. Chen and K.P. Sycara. Webmate: A personal agent for browsing and searching. In *Proceedings of the Second International Conference on Autonomous Agents*, pages 132–139, 1998.
6. P. Chesnais, M. Mucklo, and J. Sheena. The fishwrap personalized news system. In *Proceedings of the IEEE 2nd International Workshop on Community Networking Integrating Multimedia Services to the Home*, pages 275–282, 1995.
7. A. Diaz Esteban, M.J. Mana Lopez, M. de Buenaga Rodriguez, J.M. Gomez Hidalgo, and P.G. Gomez-Navarro. Using linear classifiers in the integration of user modeling and text content analysis in the personalization of a web-based spanish news service. In *Proceedings of the Workshop on User Modeling, Machine Learning and Information Retrieval, 8th International Conference on User Modeling (UM2001)*, 2001.
8. T. Kamba, H. Sakagami, and Y. Koseki. Anatagonomy: a personalized newspaper on the world wide web. *International Journal of Human-Computer Studies*, 46(6):789–803, 1997.

9. T. Kurki, S. Jokela, R. Sulonen, and M. Turpeinen. Agents in delivering personalized content based on semantic meta-data. In *Proceedings of the AAAI Spring Symposium Workshop on Intelligent Agents in Cyberspace*, pages 84–93, 1999.
10. G. Paliouras, V. Karkaletsis, C. Papatheodorou, and C.D. Spyropoulos. Exploiting learning techniques for the acquisition of user stereotypes and communities. In *Proceedings of the International Conference on User Modeling (UM), CISM Courses and Lectures, n. 407*, pages 169–178, 1999.
11. G. Paliouras, A. Mouzakidis, C. Ntoutsis, A. Alexopoulos, and C. Skourlas. Pns: Personalized multi-source news delivery. In *Proceedings of the 10th International Conference on Knowledge-Based & Intelligent Information & Engineering Systems (KES), Lecture Notes in Artificial Intelligence, n. 4252*, pages 1152–1161, 2006.
12. G. Paliouras, C. Papatheodorou, V. Karkaletsis, and C.D. Spyropoulos. Clustering the users of large web sites into communities. In *Proceedings of the International Conference on Machine Learning (ICML)*, pages 719–726, 2000.
13. G. Paliouras, C. Papatheodorou, V. Karkaletsis, and C.D. Spyropoulos. Discovering user communities on the internet using unsupervised machine learning techniques. *Interacting with Computers*, 14(6):761–791, 2003.
14. D. Pierrakos, G. Paliouras, C. Papatheodorou, and C.D. Spyropoulos. Web usage mining as a tool for personalization: a survey. *User Modeling and User-Adapted Interaction*, 13(4):311–372, 2003.
15. G. Sigletos, G. Paliouras, C.D. Spyropoulos, and M. Hatzopoulos. Combining information extraction systems using voting and stacked generalization. *Journal of Machine Learning Research*, 6:1751–1782, 2005.

Editors

Maria Virvou

Maria Virvou is an Associate Professor in the Department of Informatics, University of Piraeus, Greece. She received a degree in Mathematics from the University of Athens, Greece (1986), an M.Sc. in Computer Science from the University of London (University College London), UK (1987) and a D.Phil. from the School of Cognitive and Computing Sciences of the University of Sussex, UK (1993). She is the sole author for three computer science books. She has authored or co-authored over 200 articles, which have been published in international journals, books and conference proceedings. She has served as a member of Program Committees and/or reviewer of international journals and conferences. She has supervised or currently is supervising 12 *Ph.D.s*. She has served or is serving as the project leader and/or project member in 15 R&D projects in the areas of e-learning, computer science and information systems. She is the general co-chair and the program co-chair of the *First International Symposium on Intelligent Interactive Multimedia Systems and Services (KES-IIMSS 2008)*, organized jointly by the Department of Informatics of the University of Piraeus and KES International. Her research interests are in the areas of web-based information systems, knowledge-based human–computer interaction, personalization systems, software engineering, e-learning, e-services and m-services. She can be reached at `mvirvou@unipi.gr`.

Lakhmi C. Jain

Lakhmi C. Jain is a Director/Founder of the Knowledge-Based Intelligent Engineering Systems Centre, located at the University of South Australia, where he is Professor. He is a fellow of the Institution of Engineers Australia. His research interests focus on the novel techniques such as knowledge-based intelligent machines, artificial neural networks, virtual systems, intelligent agents, evolutionary systems and the application of these techniques. Prior to obtaining his Ph.D., he worked as a Research Fellow under the Council of Scientific and Industrial Research on the design and development of educational digital computers. His publications include more than ten co-authored books, 50 co-edited books, 15 co-edited conference proceedings, 100 research papers and ten special issues of journals. He initiated the First International Conference on Knowledge-Based Intelligent Systems in 1997. This is now an annual event. He also initiated the International Journal of Knowledge-Based Intelligent Systems in 1997. He has presented a number of Keynote addresses in International Conferences on intelligent systems and their applications. He can be reached at Lakhmi.jain@unisa.edu.au.